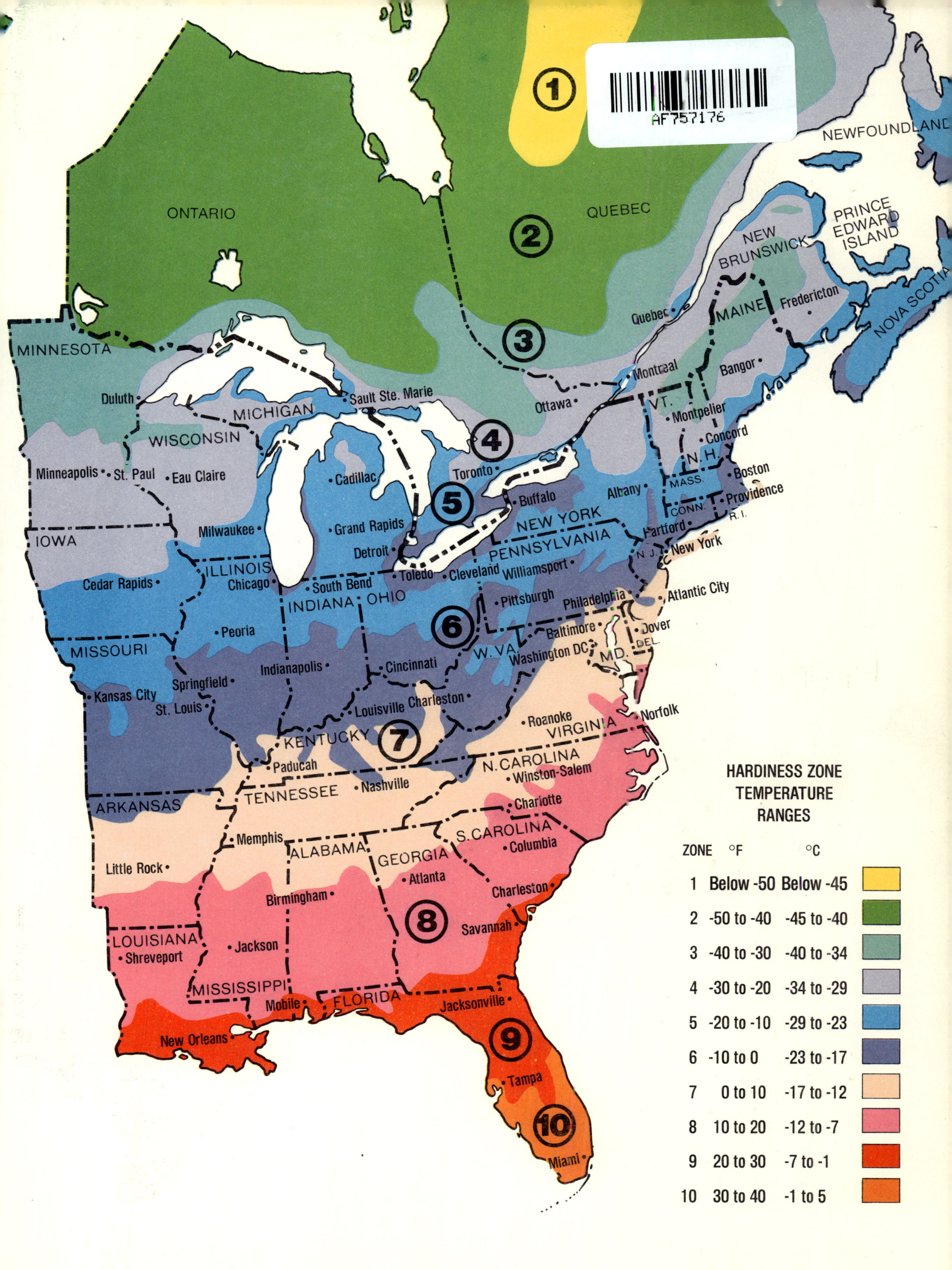

HARDINESS ZONE TEMPERATURE RANGES

ZONE	°F	°C
1	Below -50	Below -45
2	-50 to -40	-45 to -40
3	-40 to -30	-40 to -34
4	-30 to -20	-34 to -29
5	-20 to -10	-29 to -23
6	-10 to 0	-23 to -17
7	0 to 10	-17 to -12
8	10 to 20	-12 to -7
9	20 to 30	-7 to -1
10	30 to 40	-1 to 5

Pictorial Library

N
W
E
S

Pictorial Library

OF LANDSCAPE PLANTS

First Edition written by
M. Jane Coleman Helmer, Ph.D.

•

Second Edition revised and edited by
John L. Threlkeld

merchants publishing company
kalamazoo, michigan

HYBRID TEA ROSE 'COLOR MAGIC'

ISBN 0-89484-008-8

merchants publishing company • 20 mills street • kalamazoo, michigan 49001

First Edition 1974

Second Edition 1979

Printed in the United States of America

Contents

Preface

TO THE SECOND EDITION

There was no Preface or Introduction to the First Edition of the PICTORIAL LIBRARY because it was designed strictly as a sales tool for use by landscape contractors and nurserymen who needed some fine, full-page plant illustrations to show clients or prospective clients. That's all there was to the First Edition—310 pages of excellent plant illustrations—one side only, with minimal descriptive copy, loose-leaf, in a large three-ring binder.

But the PICTORIAL LIBRARY became more than a sales tool. It became a unique publication, valued by horticultural students, teachers and librarians as well as by the professionals. Many requests were received for an indexed, bound version more suited for use in public places such as in libraries and retail nurseries where the security of the pages in a hardbound book would be appreciated.

This Second Edition is designed to suit the needs of all of these horticulturally interested persons. Its materials are chosen to give it the maximum long-lasting qualities of a textbook, plus the beauty and style fit for gracing the coffee table of the most graciously decorated and appointed drawing room.

The subjects illustrated are mostly of hardy materials, chosen for their popularity in the temperate regions of North America. A few less hardy subjects are shown to somewhat broaden its scope. Subjects are printed on both sides, thus reducing the bulk, but not the content.

The book as revised remains an excellent sales tool, less bulky to carry to a sales call, and more durable and secure because of the sturdy construction and hard cover.

Dr. Helmer and I join in thanks to those who have contributed to both editions:

—Earl Klatte, Art Director in charge of design of the First Edition, and Howard F. Somer, designer of the Second Edition.

—Clarence "Clancy" E. Lewis, Ph.D., Professor Emeritus, Department of Horticulture, Michigan State University, East Lansing, Michigan, for technical consultations.

—John Pike and the late Willard Kalina for the horticultural photography.

KALAMAZOO, MI. JOHN L. THRELKELD
APRIL, 1979

Introduction

HOW TO USE THIS BOOK

The design and layout of this book is simple and straightforward. It is to illustrate ornamental and fruiting plants in typical landscape settings, to show the characteristics of the plants for identification as well as for visual appreciation in a typical use. The book is divided into ten sections or categories of plants, alphabetically arranged in each section by botanical names. If you are interested in a plant and know in which section it falls, simply turn to that section and locate it alphabetically. If you don't know the section, turn to the index which is arranged alphabetically by both botanical and common names.

EXPLANATION OF THE SECTIONS

Broadleaf Evergreen Shrubs: These are the bold, usually showy shrubs excellent for use as foundation plantings for energy conservation and beautification. Also valued in shrub borders and as background, screen and hedging.

Coniferous Evergreens: This section includes shrubs and trees of the conifer (cone-producing) or needle-leaf types. These are non-flowering, but they are always attractive and are the basic foundation and windbreak plants for energy conservation and beautification. These are popularly referred to as "evergreens."

Deciduous Shrubs: Most of the brilliantly-colored, flowering accent shrubs are located here. Many of these are often used as foundation plants, but they don't offer the wintertime protection afforded by the evergreens.

Flowering Trees: These are tree types valued for color accents and maximum ornamental effect. They provide excellent bird shelters. Most of these trees can be used on small lots and in confined areas such as patios and dooryard gardens.

Shade Trees: Here are shown the larger deciduous trees, valued for their cooling summer shade and their leafless nature in winter that admits warming sunshine. Some of these specimens grow very large in time, and care must be taken in their placement to assure against future overgrowth that can become a serious maintenance problem.

Food-Producing Plants: With the mounting popularity of citizens growing much of their own food in gardens and orchards, it is appropriate to illustrate some of the more popular types here with some cultural directions.

Ground Covers: Here are to be seen some examples of landscape uses of plants to both beautify and reduce maintenance labor by covering the ground totally, thus reducing weeds and erosion problems. Some of both of the broadleaf evergreen and coniferous evergreen types are illustrated.

Roses: These popular plants need no introduction except to note the different types, from the shrub and border types to the climbing and rambling kinds.

Hedges: While there is some duplication here, it is important to illustrate several different ideas for hedging and screening, as these are such important segments of the landscape.

Vines: Some of the most colorful and effective landscape features can be achieved with vines. Vines are too often overlooked in landscape planning, and some of these gaudy beauties are shown here to serve as a reminder of their great value.

ORDER OF PLANT NAMES

Leading in bold type is the most universally accepted common name; but all arrangement is alphabetical by botanical name. This is because different regions use different common names for the same plant, and to alphabetize by common name would lead only to confusion. If any confusion results in trying to locate a plant when you know only a common name, please refer to the Index.

To help understand the order used here, the following examples are given:

BOXWOOD. . . . . common name
Buxus. . . . . genus name
Buxus microphylla. . . . . the species is the second name.
Buxus microphylla var. koreana. . . . . the third name if any, is the natural variety.
Buxus microphylla var. koreana 'Wintergreen'. the fourth name in half-quotes is the cultivar (formerly, variety, or a plant developed by man). A cultivar may appear in any position after the genus as with a hybrid: Malus hybrid 'Red Silver' or as a specific cultivar: Picea pungens 'Glauca', and is always in half-quotes.
(Thuja orientalis). . . . . a synonym
Abelia ×grandiflora. . . . . times sign preceding the species indicates hybrid species; preceding the genus indicates hybrid genus.

Abbreviations (used mostly in the index):

cv. . . . . cultivar
sp. . . . . species singular
spp. . . . . species plural
var. . . . . natural variety

Broadleaf Evergreens

GOLD DUST PLANT *Aucuba japonica 'Variegata'*

Zones 6 7 8 9 10

Evergreen shrub, height to 6-15′ with equal spread in 10-15 years.

Flowers inconspicuous, flower sexes separate.

Fruit: brilliant red berries ripen in fall and persist through winter, needs both sexes to insure fruits.

Leaves up to 6″ long, thick and leathery; glossy dark green with yellow speckles.

Soil: moist and well drained.

Light to deep shade.

TOLERATES drought once established.

Pruning: rarely needed. Clip individual branches as desired in early spring.

Uses: container, patio, indoor plant, shrub borders, woodland areas.

Broadleaf Evergreens

GLOSSY ABELIA *Abelia ×grandiflora*

Zones 6 7 8 9 10

Semi-evergreen dense shrub. Height and spread 3-5′ in 4-5 years. Severe winters cause leaf fall. All top growth may be killed back by very low temperatures (0°F), but plant will grow from ground level to flower by midsummer.

Flowers on current season's growth, from June through fall; clusters of 1″ long tubular rosy-white blossoms.

Leaves 1″ long; reddish in spring turning glossy dark green in summer.

FALL COLOR: glossy bronze purple, leaves held through milder winters.

Soil: almost any, well drained.

Light: full sun or light shade; shelter from winds.

Prune in early spring. Hedges may be sheared, though prefer to clip out individual stems.

Uses: foundation planting, formal or informal hedge, shrub border.

JAPANESE AUCUBA *Aucuba japonica*

Zones 6 7 8 9 10

Evergreen shrub, height and spread 6-15′ in 10-20 years. Plant both male and female shrubs for good fruit production.

Flowers inconspicuous; small white blossoms in early spring.

Fruit: brilliant red berries ripen in fall and persist through winter.

Leaves up to 6″ long, glossy dark green with thick and leathery texture.

Soil: almost any; best in moist and well drained position.

TOLERATES drought once established.

Light to deep shade.

Pruning rarely needed. If desired, clip individual branches in early spring.

Uses: container plant for patio and indoor use, shrub border, woodland garden.

GOLD DUST PLANT *Aucuba japonica 'Variegata'*

Zones 6 7 8 9 10

Evergreen shrub, height to 6-15′ with equal spread in 10-15 years.

Flowers inconspicuous, flower sexes separate.

Fruit: brilliant red berries ripen in fall and persist through winter, needs both sexes to insure fruits.

Leaves up to 6″ long, thick and leathery; glossy dark green with yellow speckles.

Soil: moist and well drained.

Light to deep shade.

TOLERATES drought once established.

Pruning: rarely needed. Clip individual branches as desired in early spring.

Uses: container, patio, indoor plant, shrub borders, woodland areas.

DARWIN'S BARBERRY *Berberis darwinii*

Zones 6 7 8 9 10

Jpright semi-evergreen shrub, evergreen in milder climates. f unpruned, height reaches 6′ with spread of about 5′. Three-pronged spines along stems.

'lowers line stems in spring, ¼″ yellow blossoms. 'ollowed by blue to black berries which ersist into winter.

eaves 1-3″ long, glossy green, spiny edges.

oil: almost any, well drained; tolerates lime.

Light: full sun to light shade.

TOLERATES heat and drought, but not wind.

Prune to shape; shear several times each season to maintain dense, formal appearance. Specimen plants rarely need pruning except to contain irregular growth.

Uses: hedge, shrub border, specimen or accent.

KOREAN BOXWOOD *Buxus microphylla var. koreana 'Wintergreen'*

Zones 6 7 8 9 10

Low evergreen shrub; height to 2′, spread to 4′ in 8-10 years. Best established when planted in spring.

Flowers in spring, small creamy-white clusters at ends of branches.

Leaves ½-1″ ovals, dark glossy green retained throughout winter.

Soil: almost any, moist and well drained. Mulch with 1″ layer or coarse peatmoss, wood chips or peanut shells to retain moisture in summer. Syringe during dry spells until established (1-2 years).

Light: full sun or light shade. Protect from winter winds with burlap or other screening.

Prune only to remove dead branches. Shear formal hedges.

Uses: formal or informal hedge; plant at 12-18″ and remove tips of growing shoots to encourage bushy growth.

BOXWOOD WELLERII *Buxus sempervirens 'Wellerii'*

Zones 6 7 8 9 10

Wide spreading evergreen shrub, height 2-5′ and spread 5-10′ in about 15-20 years. Best established when planted in spring.

Flowers in spring, small white clusters at ends of branches.

Fruit: shiny black persistant berries.

Leaves: 1-1½″ ovals with delicate fragrance especially in warmer climates. Cold winters causes bronzing.

Soil: almost any, moist and well drained. Mulch with 1″ layer of coarse peatmoss, wood chips or peanut shells to retain moisture in summer. Syringe plants during dry spells until established (1-2 years).

Light: full sun or light shade. Protect from winter winds with burlap or other screening.

Prune only to remove dead branches. Shear formal hedges.

Uses: formal or informal hedge; plant at 15-24″ and remove tips of growing shoots to encourage bushy growth.

HEATHER *Calluna vulgaris & varieties*

Zones 4 5 6 7 8

Spreading evergreen plants, heights range from 8″ to 24″.

Flowers from late summer into fall, tiny blossoms in spikes at shoot tips. Varieties in white, pinks and purples.

Leaves tiny and scale-like, dark green (some varieties have gold-colored foliage)

Soil: low fertility, sandy and well drained; prefers acid conditions. Shallow root system.

Light: full sun; tolerates some shade though with fewer flowers.

Prune: shear back to half height in early spring just before new growth begins.

Uses: ground cover, front of borders, low hedge or edging, rock garden, patio or container plants.

THORNY ELAEAGNUS *Elaeagnus pungens*

Zones 6 7 8 9 10

Evergreen spreading shrub, growing to 8-15′ in 8-10 years. Easily transplanted.

Flowers in late fall, small fragrant ivory-white blossoms.

Fruit mature by spring, ½″ red berries.

Leaves 2-4″ long with undulating margins, dark green color all year; silvery-tan undersides.

Soil: almost any, wet or dry.

Light: full sun or light shade.

TOLERATES heat and wind.

Prune at any time; shear hedge or clip out individual stems.

Uses: screen, hedge, in shrub borders.

VARIEGATED THORNY ELAEAGNUS *Elaeagnus pungens 'Maculatus'*

Zones 6 7 8 9 10

Evergreen spreading shrub, growing to 8-15′ in 8-10 years. Easily transplanted.

Flowers in late fall, small fragrant ivory-white blossoms.

Fruit mature by spring, ½″ red berries.

Leaves 2-4″ long with undulating margins; large gold colored blotch in center of dark green, color holds all year.

Soil: almost any, wet or dry.

Light: full sun or light shade.

TOLERATES heat and wind.

Prune at any time; shear hedge or clip out individual stems.

Uses: screen, hedge, in shrub borders.

SPRING HEATH *Erica carnea*

Zones 5 6 7 8

Spreading evergreen shrub, height 6-12″.

Flowers from late fall in mild climates, early spring in colder areas. White, pink, purple or red blossoms for about three months.

Leaves dark green and needle-like.

Soil: almost any, moist and well-drained; prefers acid conditions. Shallow root system.

Light: best in full sun.

Prune: for bushy growth, shear back to about half height after flowering.

Uses: ground cover, front of borders, rock gardens.

WINTERCREEPER *Euonymus fortunei*

Zones 5 6 7 8 9 10

Evergreen spreading creeper, erect stems to height of 3-4′ (unsupported), spreading 4-6′ in 8-10 years.

Leaves: ½-2″ ovals, glossy dark green. Holds leaves through all but severest winters.

Soil: almost any, wet or dry; well drained.

Light: full sun or light shade

Prune at any time; responds well to shearing.

Uses: low edging, ground cover, foundation planting.

SARCOXIE EUONYMUS *Euonymus fortunei 'Sarcoxie'*

Zones 5 6 7 8 9 10

Evergreen shrub, upright and spreading. Height to 4′ in 8-10 years.

Leaves 1″ long, glossy dark green all year.

Fruit: orange red, resemble bittersweet.

Soil: almost any, wet or dry.

Light: full sun or light shade.

Prune at any time to shape; responds well to shearing.

Uses: low hedge, foundation planting.

BIGLEAF WINTERCREEPER or EVERGREEN BITTERSWEET *Euonymus fortunei var. vegeta*

Zones 5 6 7 8 9 10

Evergreen shrubby spreader, erect stems to height of 4′, spreads to about 4′ in 8-10 years.

Fruit: orange red, resembles bittersweet.

Leaves 1-1½″ rounded ovals, glossy dark green all year.

Soil: almost any, wet or dry.

Light: full sun or shade.

Prune at any time to shape; responds well to shearing.

Uses: low hedge, ground cover, foundation planting, train on support: espalier.

EMERALD 'N GOLD *Euonymus fortunei var. vegeta 'Emerald 'n Gold'*

Plant Patent #2231

Zones 5 6 7 8 9 10

Evergreen shrubby spreader, erect stems to height of 4′, spreads to about 4′ in 8-10 years.

Leaves 1-1½″ rounded ovals, glossy dark green irregular centers with golden yellow margins.

Soil: almost any, wet or dry.

Light: full sun or light shade.

Prune at any time to shape; responds well to shearing.

Uses: low hedge, ground cover, foundation planting.

PEARL EDGE VARIEGATED EVERGREEN EUONYMUS *Euonymus japonica 'Albo-marginata'*

Zones 6 7 8 9 10

Evergreen shrub or small tree growing to height of 6-10′ in 8-10 years if unpruned. Easy to shape and maintain smaller size.

Leaves 3″ long, glossy dark green with thin margin of white; color held throughout year.

Soil: almost any, wet or dry.

Light: full sun or light shade.

Prune at any time to shape; responds well to shearing.

Uses: hedge or screen, accent, shrub border.

GOLDEN EUONYMUS *Euonymus japonica 'Aureo-marginata'*

Zones 6 7 8 9 10

Evergreen shrub or small tree growing to height of 6-10′ in 8-10 years if unpruned. Easy to shape and maintain smaller size.

Leaves 3″ long, glossy dark green with yellow edges; color held throughout year.

Soil: almost any, wet or dry.

Light: full sun or light shade.

Prune at any time to shape; responds well to shearing.

Uses: hedge or screen, accent, shrub border.

EUONYMUS PATENS *Euonymus kiautschovicus*

Zones 6 7 8 9 10

Spreading semi-evergreen shrub, evergreen in milder climates. Height and spread to 4-5′ in 5-6 years. Flowers in September.

Fruit develops late in year, brilliant red and white, in clusters.

Leaves: glossy dark green, hold color all year in milder climates.

Soil: almost any, well drained.

Light: full sun or light shade.

Pruning: rarely needed.

Uses: specimen or accent, group planting.

CAPE JASMINE or GARDENIA

Gardenia jasminoides 'Mystery' or 'Miami Supreme'

Zone 9, plus southeast zone 8 and southern area in zone 10

Evergreen flowering shrub, reaching height of 4-6′. Late spring frosts will damage flowers and young leaves.

Flowers about 3″ across, white and waxy, very fragrant. Buds form when night temperatures are less than 65°, but flowering ceases below 55°F.

Leaves about 4″ long, thick and leathery; glossy dark green color retained all year.

Soil: moist and acid, prefers lime-free conditions. Supplement with peatmoss, leafmold or compost (1 part to 2 parts soil, by volume). Use permanent mulch 2-3″ deep. Surface rooted—avoid cultivating around the plant.

Light: full sun or light shade.

Prune in early spring to remove dead wood and to shape plant.

Uses: accent, container plant for patio and indoor use.

CHINESE HOLLY *Ilex cornuta*

Zones 6 7 8 9 10

Evergreen shrub; height to 6-10′ in 5-10 years; spreads 6-8′. Flowers and fruits on previous season's growth.

Fruit: bright red, in large clusters.

Leaves 3-4″ long, very glossy dark green, spines at corners and tip.

Soil: almost any, prefers slightly acid conditions. Tolerates moist position. Mulch rather than cultivate around plants.

Light: best in full sun; tolerates shade though bears fewer berries.

Pruning rarely needed; trim hedge in summer.

Uses: specimen or accent, hedge.

BURFORD HOLLY *Ilex cornuta 'Burfordii'*

Zones 6 7 8 9 10

Vigorous upright and compact evergreen shrub. Height 6-10′ in 8-12 years. Flowers and fruits on previous season's growth.

Fruit: prolific red berries.

Leaves 4′ long, few or no spines, glossy dark green.

Soil: almost any, prefers slightly acid conditions. Tolerates moist position. Mulch rather than cultivate around plants.

Light: best in full sun; tolerates shade though with fewer berries.

Pruning rarely needed; trim hedge in summer.

Uses: specimen or accent, hedge.

DWARF BURFORD HOLLY *Ilex cornuta 'Burfordii Nana'*

Zones 6 7 8 9 10

Compact evergreen shrub. Slow growing, mature height and spread 3-4′. Flowers on previous season's growth.

Fruit: prolific red berries.

Leaves 3-4″ long with few if any spines; glossy dark green.

Soil: almost any, prefers slightly acid conditions. Tolerates moist position. Mulch rather than cultivate around plants.

Light: best in full sun. Tolerates shade but produces fewer berries.

Pruning: rarely needed. Trim hedge in summer.

Uses: low hedge, foundation planting, accent in small area.

DWARF CHINESE HOLLY *Ilex cornuta 'Rotunda'*

Zones 6 7 8 9 10

Low growing compact evergreen shrub with densely branching habit.

Sterile—no berries produced.

Leaves very glossy dark green, spines at corners and tip.

Soil: almost any, prefers slightly acid conditions.

Tolerates moist position. Mulch rather than cultivate around plants.

Light: full sun or light shade.

Pruning rarely needed; trim hedge in summer.

Uses: low hedges, edge for borders, flower beds.

JAPANESE HOLLY *Ilex crenata 'Convexa'*

Zones 6 7 8 9 10

Upright and bushy evergreen shrub. Height 8-10′ in about 10-15 years if unpruned. Inconspicuous flowers, black fruit.

Leaves 1″ long, finely tooth, dark green.

Soil: almost any, prefers slightly acid conditions. Tolerates moist position. Mulch rather than cultivate around plants.

Light: full sun or light shade.

Pruning rarely needed; shear to desired size.

Uses: hedge or screen.

DWARF JAPANESE HOLLY *Ilex crenata 'Compacta'*

Zones 6 7 8 9 10

Compact, profusely branching evergreen shrub. Height 12″, spread 36″ in about 10 years. Inconspicuous flowers, black fruit.

Leaves ½-1″ long, finely toothed, dark green.

Soil: almost any, prefers slightly acid conditions. Tolerates moist position. Mulch rather than cultivate around plants.

Light: full sun or light shade

Pruning: rarely needed; trim as desired.

Uses: low hedge or edging, container or patio plant, ground cover.

NORTHERN BEAUTY HOLLY *Ilex crenata 'Convexa Northern Beauty'* Zones 6 7 8 9 10

Low spreading evergreen shrub. Height 3-4′ in about 10-15 years, spreads to 6′ if unpruned. Inconspicuous flowers and black fruit.

Leaves ½-1″ long, convex, glossy dark green.

Soil: almost any, prefers slightly acid conditions. Tolerates moist position. Mulch rather than cultivate around plants.

Light: best in full sun.

Pruning rarely needed. Trim hedge in summer.

Uses: low hedges, ideal for edging or below windows, foundation planting.

HELLER'S JAPANESE HOLLY *Ilex crenata 'Helleri'*

Zones 6 7 8 9 10

Dwarf evergreen shrub with height 12-18″ and spread 36″ in 8-10 years. Prostrate and very compact growth, looks naturally sheared. Inconspicuous flowers, black fruit.

Leaves ½″ ovals, glossy dark green.

Soil: almost any, prefers slightly acid conditions. Tolerates moist position. Mulch rather than cultivate around plants.

Light: full sun or light shade.

Pruning: rarely necessary.

Uses: low hedge or edging, container or patio plant, ground cover.

AMERICAN HOLLY *Ilex opaca*

Zones 6 7 8 9 10

Broadly conical evergreen shrub. Height 8-15′ in 10-15 years, matures to 30-40′ if unpruned. Need both male and female plants for good fruit production.

Flowers on current season's growth, in summer.

Fruit: abundant red berries, rarely eaten by birds, persist through winter.

Leaves spiny, dull green with yellow-green undersides.

Soil: almost any, prefers slightly acid conditions. Tolerates moist position. Mulch rather than cultivate around plants.

Light: best in full sun; tolerates shade though bears fewer berries.

Pruning: rarely needed.

Uses: specimen or accent.

YAUPON HOLLY *Ilex vomitoria*

Zones 7 8 9 10

Multi-stemmed evergreen shrub. Unpruned height and spread to 15-20′ in 15-20 years. Easy to shear to any height. Flowers and fruits on previous season's growth. Need both male and female plants for good fruit production.

Fruit: abundant translucent red berries.

Leaves 1″ long, gray-green with smooth margins.

Soil: almost any, prefers slightly acid conditions. Tolerates moist position. Mulch rather than cultivate around plants.

Light: best in full sun; tolerates shade though bears fewer berries.

Pruning: rarely needed; shear to desired size.

Uses: hedge, up to 12′ tall, large specimen or accent.

DWARF YAUPON HOLLY *Ilex vomitoria 'Nana'*

Zones 7 8 9 10

Dwarf multi-stemmed evergreen shrub, growing to height and spread of 18-48″. No fruit.

Leaves 1″ long, gray-green with smooth margins.

Soil: almost any, prefers slightly acid conditions. Tolerates moist position. Mulch rather than cultivate around plant.

Light: full sun or light shade.

Pruning: rarely needed; trim hedge in summer.

Uses: ground cover, container, patio, low hedge.

OREGON GRAPE *Mahonia aquifolium*

Zones 5 6 7 8 9 10

Evergreen shrub growing to height and spread of 3-6′ in 5-6 years.

Flowers in spring, short clusters of bright yellow blossoms at ends of stems.

Fruit: blue-black and grape-like in appearance.

Leaves typical holly leaf shaped leaflets in groups of 5, 7 or 9. Spiny and leathery, dark green turning bronze in cold weather. Bronze-green when young.

Soil: moist, best in acid conditions; add one third part peatmoss leafmold or compost to soil. Will tolerate neutral or alkaline soils.

Light shade.

TOLERATES heat, drought and cold.

Prune tall canes (stems) to ground in early spring for a succession of vigorous young shoots.

Uses: foundation plantings, in shrub borders, specimen or accent.

COMPACT OREGON GRAPE *Mahonia aquifolium 'Compacta'* Zones 5 6 7 8 9 10

Low growing evergreen shrub, grows to height of 18-30″.

Flowers in spring, dense clusters of bright yellow blossoms.

Fruit: blue-black and grapelike.

Leaves: typical holly-leaf shaped leaflets in groups of 5, 7 or 9. Spiny and leathery textured, very glossy. Light bronze-green when young, maturing dark green and turning bronze in fall and winter.

Soil: moist, best in acid soil; add peatmoss, leafmold or compost. Will tolerate neutral or alkaline conditions.

Light: best in shade.

TOLERATES heat, drought and cold.

Prune older stems to the ground in early spring to encourage succession of vigorous young shoots.

Uses: ground cover, foundation plantings, shrub borders.

HEAVENLY BAMBOO *Nandina domestica*

Zones 6 7 8 9 10

Evergreen shrub growing to height of 6′, spread of 4′. Plant several to encourage cross pollination for maximum fruit production.

Flowers in July, small white blossoms in 6-12″ clusters.

Fruit: small dark red berries in clusters, mature in fall, persist into winter.

Leaves bronze red when young, turning deep green in summer.

FALL COLOR: red to scarlet, holds all winter.

Soil: moist and well drained; add one third part peatmoss, leafmold or compost to soil.

Light: full sun or light shade.

Prune in early spring to cut older or overgrown stems to the ground.

Uses: foundation planting, shrub borders, accent or specimen.

OLEANDER *Nerium oleander*

Zones 8 9 10

Evergreen shrub growing to height and spread of 8-10′ in about 8 years. New growth will start from the ground if top is killed in severe winter.

Flowers from April through summer; 1-2″ single or double blossoms, often fragrant varieties in yellow, white, red, salmon or pink.

Leaves 4-6″ long and narrow, green all year.

Soil: almost any, moist and well drained; add equal volume of peatmoss or leafmold to garden soil.

Light: full sun.

TOLERATES heat, drought, urban pollution and coastal climate.

Prune to encourage branching by cutting off stem tips after flowering. Plants may be cut back hard in spring if required.

DO NOT PRUNE unless you sterilize the shears after each cut. Use formaldehyde or chlorine, 1 oz. per qt. of water. This is to discourage spread of gall disease.

Uses: specimen or accent, roadside planting, informal hedge, screen, windbreak, patio or container plant.

JAPANESE PHOTINIA *Photinia glabra*

Zones 6 7 8 9 10

Evergreen shrub growing to 10-12′ if unpruned, in 10-15 years. Spread equals or exceeds height.

Flowers in spring, 4″ clusters of white blossoms.

Fruit ripen in fall, red berries.

Leaves 3-6″ long with toothed margins; bronze or red when young, maturing to glossy dark green.

Soil: almost any, well drained; add equal volume of peatmoss or leafmold to garden soil.

Light: full sun or light shade.

Prune to control size and encourage branching. Pinch off growing tips every time new shoots are 18″ long.

Uses: specimen or accent, shrub border, hedge.

JAPANESE ANDROMEDA *Pieris japonica*

Zones 5 6 7 8, plus north in zone 10

Evergreen flowering shrub growing to height and spread of 4-8′ in 10-12 years.

Flowers in April and May, 3-6″ long drooping clusters of white or blush white blossoms.

Leaves: densely packed 1½-3″ oblong. Rich bronze when young in spring, maturing glossy dark green in summer; color retained all year.

Soil: moist and acid; add peatmoss, leafmold or compost. Permanent mulch 3-4″ deep will retain moisture and keep roots cool.

Light: best in partial shade especially in hot dry periods. Tolerates sun if soil is moist.

Prune to shape and to remove dry seed capsules right after flowering.

Uses: specimen, acid shrub borders, foundation planting.

JAPANESE PITTOSPORUM *Pittosporum tobira*

Zones 8 9 10,

Evergreen dense shrub, maturing to 8-10′ high with equal or greater spread. Reaches 6′ in 8-10 years.

Flowers in spring, 2-3″ clusters of creamy white blossoms; fragrance reminiscent of orange blossom.

Leaves 2-4″ long, thick and rubbery; dark glossy green.

Soil: almost any, well drained.

Light shade.

TOLERATES heat and coastal windy climates; also withstands drought once established.

Prune with hand clippers to shape as desired.

Uses: hedges and screens, foundation planting, shrub border.

VARIEGATED JAPANESE PITTOSPORUM *Pittosporum tobira 'Variegata'* Zones 8 9 10

Evergreen shrub growing to height of 4-6′; compact and dense habit.

Flowers in May; ½″ creamy white blossoms in clusters; fragrance reminiscent of orange blossom.

Leaves 2-4″ long, thick and rubbery; pale green or gray green with white margins.

Soil: almost any, well drained.

Light shade.

TOLERATES heat and coastal climates; also withstands drought once established.

Prune with hand clippers to shape as desired.

Uses: shrub border, foundation planting, low hedges.

SC ... ET FIRE THORN *Pyrancantha coccinea 'Lalandii'*

Zones 5 6 7 8 9 10

Upright ev... shrub. Fast growing. Height 6-9′ in 5-8 years. Stiff ...s with spiny thorn-like spurs.

Flowers in s... lusters of tiny white blossoms.

Fruit: shiny ... ed in clusters, persist through winter if not t... birds.

Leaves: small p...ited ovals, dark green and glossy.

Soil: almost any, well drained.

Light: full sun.

Prune at any time; clip off ends of growing shoots during growing season to encourage branching.

Uses: barrier, trained against wall or fence—espalier, hedge, background.

GRABERI FIRE THORN *Pyrancantha fortuneana 'Graberi'*

Zones 7 8 9 10

Erect vigorous evergreen. Height 7-9′ in 5-8 years.

Flowers in spring, clusters of tiny white blossoms.

Fruit: huge clusters of bright red berries appear early, in midsummer; persist until following spring if not taken by birds.

Leaves: small pointed ovals, dark green and glossy.

Soil: almost any, well drained.

Light: full sun.

Prune at any time; clip off ends of long shoots during growing season to encourage branching.

Uses: trained against wall or fence—espalier, background, container grown for patios, accent.

ESPALIERED FIRETHORN *Pyracantha species*

Zones 5 6 7 8 9 10

Vigorous evergreen shrubs, growing as much as 9′ in 5-8 years. Stiff branches with spiny thornlike spurs. Can be trained to height of 6-18′.

Flowers in spring, clusters of tiny white blossoms.

Fruit: clusters of shiny red or orange-red berries ripen in midsummer or fall. Birds love them.

Leaves: small pointed ovals, dark green and glossy.

Soil: almost any, well drained.

Light: full sun.

Prune at any time; clip off the ends of long shoots during the growing season to encourage branching.

Train by tying young pliable branches to wall or other support, using loose ties of soft string, raffia or rubber budding bands (not wire). Loose ties permit healthy development of the tied stem; check each year to be sure the thickening stem has not become constricted by its tie, and replace any ties that are tight. For a true espaliered look, plan to space the main branches 15-18″ apart. Prune out unwanted shoots at any time.

INDICA AZALEA *Rhododendron indicum—Southern Indica Group*

Zones 8 9 10

Evergreen flowering shrub; height to 6′, spread equals or exceeds height.

Flowers in late June, 2-3″ blossoms.

Leaves 1-2″ long, mid to dark green color all year.

Soil: moist and well drained, acid (pH 4.5-5.5). Add peatmoss, leafmold or compost. Surface root system; mulch annually to retain moisture. Water well when in flower.

Light shade; protect from heat of midday and afternoon sun. Deep shade or continual damp will encourage mildew attack. Protect from winds.

Pruning rarely needed; remove dead and unsightly stems.

Uses: foundation planting, shrub border, light woodland, accent.

HINODEGIRI AZALEA *Rhododendron obtusum 'Hinodegiri'*

Zones 6 7 8 9 10

Evergreen flowering shrub, semi-evergreen in colder climates. Height to 5′; spread equals or exceeds height.

Flowers in May, 1½″ red blossoms.

Leaves mid to dark green all summer.

Soil: moist and well drained, acid (pH 4.5-5.5); add peatmoss, leafmold or compost. Surface root system. Water well when in flower; mulch annually to retain moisture.

Light shade; protect from heat of midday and afternoon sun. Protect from winds at all times.

Pruning rarely needed; remove dead and unsightly stems.

Uses: foundation planting, shrub borders, informal or formal hedge, light woodland.

KAEMPFERI AZALEA *Rhododendron obtusum var. kaempferi*

Zones 6 7 8 9 10

Evergreen flowering shrub, semi-evergreen in colder climates. Height to 5′; spread equals or exceeds height.

Flowers in mid-May; 2-3″ salmon to rose pink blossoms.

Leaves mid to dark green all summer, brownish in fall or winter.

Soil: moist and well drained, acid (pH 4.5-5.5); add peatmoss, leafmold or compost. Surface root system. Water well especially if dry during flowering. Mulch annually to retain moisture in summer.

Light shade; protect from heat of midday and afternoon sun. Protect from winds.

Pruning rarely needed; remove dead or unsightly stems.

Uses: foundation planting, shrub borders, light woodland, accent.

KURUME HYBRID AZALEAS *Rhododendron obtusum Kurume Group*

Zones 7 8 9 10

Semi-evergreen to evergreen flowering shrubs. Height to 5′; spread equals or exceeds height. Dense growing.

Flowers in late spring; 1½-2½″ blossoms in white, pinks, reds or oranges.

Leaves mid to dark green all summer, brownish to bronze in fall.

Soil: moist and well drained, acid (pH 4.5-5.5); add peatmoss, leafmold or compost. Surface root system.

Water well especially if dry during flowering. Mulch annually to retain moisture in summer.

Light shade: protect from heat of midday and afternoon sun. Protect from winds.

Pruning rarely needed; remove dead or unsightly stems.

Uses: foundation planting, shrub borders, light woodland, accent, flowering hedge or screen.

RHODODENDRON *Rhododendron species and hybrids*

Zones 6 7 8
and north in 10.

Evergreen shrubs with sizes ranging from 3-4′ to 15-20′ height and spread. Variable hardiness, may be classified as H-1 (hardy to minus 25°F) through H-7 (hardy to 32°F).

Flowers in clusters at stem ends; colors range from white to pinks, reds, purples, yellow, or bi-colors.

Leaves glossy dark green, often with white or tan-colored hairs on undersides.

Soil: moist and well drained, acid (pH4.5-5.5); add peatmoss, leafmold or compost. Surface root system. Water well especially during hot dry spells. Mulch annually to retain moisture in summer.

Light shade; protect from drying winds all year.

Prune individual stems rather than overall shearing to shape. Remove faded flowers or seed capsules, cutting off only the shoot tips since leaf buds immediately below the flowers will develop for next year's flowers.

Uses: foundation planting, light woodland areas, shrub border, screens, small or large, accent or specimen.

RHODODENDRON *Rhododendron species and hybrids*

Zones 6 7 8

Evergreen shrubs with sizes ranging from 3-4′ to 15-20′ height and spread. Variable hardiness, may be classified as H-1 (hardy to minus 25°F) through H-7 (hardy to 32°F).

Flowers in clusters at stem ends; colors range from white to pinks, reds, purples, yellow, or bi-colors.

Leaves glossy dark green, often with white or tan-colored hairs on undersides.

Soil: moist and well drained, acid (pH4.5-5.5); add peatmoss, leafmold or compost. Surface root system. Water well especially during hot dry spells. Mulch annually to retain moisture in summer.

Light shade; protect from drying winds all year.

Prune individual stems rather than overall shearing to shape. Remove faded flowers or seed capsules, cutting off only the shoot tips since leaf buds immediately below the flowers will develop for next year's flowers.

Uses: foundation planting, light woodland areas, shrub border, screens, small or large, accent or specimen.

RHODODENDRON *Rhododendron species and hybrids*

Zones 6 7 8
and north in 10.

Evergreen shrubs with sizes ranging from 3-4′ to 15-20′ height and spread. Variable hardiness, may be classified as H-1 (hardy to minus 25°F) through H-7 (hardy to 32°F).

Flowers in clusters at stem ends; colors range from white to pinks, reds, purples, yellow, or bi-colors.

Leaves glossy dark green, often with white or tan-colored hairs on undersides.

Soil: moist and well drained, acid (pH4.5-5.5); add peatmoss, leafmold or compost. Surface root system. Water well especially during hot dry spells. Mulch annually to retain moisture in summer.

Light shade; protect from drying winds all year.

Prune individual stems rather than overall shearing to shape. Remove faded flowers or seed capsules, cutting off only the shoot tips since leaf buds immediately below the flowers will develop for next year's flowers.

Uses: foundation planting, light woodland areas, shrub border, screens, small or large, accent or specimen.

LEATHERLEAF VIBURNUM *Viburnum rhytidophyllum*

Zones 5 6 7 8 9 10
(not southern Florida)

Evergreen shrub maturing 7-10′ tall.

Flowers in late spring, creamy white fertile flowers in 4-8″ clusters.

Fruit: red berries turning black in fall. Eaten by birds.

Leaves 3-7″ long, narrow. Dark green, deeply veined and leathery, silver green undersides.

Soil: almost any, well drained; tolerates moist conditions.

Light: full sun or light shade; tolerates shade.

Prune only to remove dead or weak shoots. Shear to guide growth.

Uses: specimen, in foundation planting, hedge, plant 3-6′ apart.

Coniferous Evergreens

WHITE FIR *Abies concolor*

Zones 4 5 6 7, plus north and west in zone 8.

Narrow pyramidal evergreen tree. Height reaches 50′ in 30-50 years. Matures to 100′.

Needles: about 2″ long. Bluish green aromatic and soft textures. Color holds all year. Erect 3-5″ cones on upper branches of older trees; Color changes from green to purplish green.

Soil: Almost any, moist and well drained, prefers acid soils.

TOLERATES urban pollution, heat and drought.

Light: Best in full sun; young plants tolerate some shade.

Pruning: Rarely required.

Uses: specimen or accent.

ATLAS CEDAR *Cedrus atlantica*

Zones 6 7 8 9 10

Fast growing broad pyramidal tree; branches curve upwards. Height can reach 40-60′ in 30-50 years. Tree becomes flat topped in maturity. Not easy to transplant.

Needles 1-2″ long, in clusters. Silvery or light green color all year.

Cones on upper branches, erect, 2-5″ high and 2″ across.

Soil: almost any, well drained.

Light: full sun.

Pruning rarely needed. If desired, clip individual branches to shape young tree.

Uses: accent tree.

DEODAR CEDAR *Cedrus deodara*

Zones 7 8 9 10

Fast growing broadly conical evergreen tree; horizontal branches with pendulous tips. Can grow 2′ each year when young, matures to 40-60′ in 30-40 years. Not easy to transplant.

Needles soft bluish green color all year.

Cones erect on branches, 4″ long and 2½″ across.

Soil: almost any, well drained.

Light: full sun.

Pruning: rarely needed.

Uses: specimen or accent.

DWARF HINOKI FALSE CYPRESS *Chamaecyparis obtusa 'Nana Gracilis'*

Zones 5 6 7 8
and north in 10

Slow growing dwarf shrub, reaches 4′ height with 3′ spread in 10-12 years. Best in cooler climates.

Foliage deep glowing green.

Soil: fertile, moist and well drained.

Light: full sun or light shade.

Pruning: rarely needed; shear in early spring to shape plant if desired.

Uses: bonsai, rock garden, patio, low hedge.

GOLDEN THREAD SAWARA FALSE CYPRESS *Chamaecyparis pisifera 'Filifera Aurea'*

Zones 5 6 7 8
and north in 10

Low shrub or small tree with drooping stringy branches; forms dense mound when young. Best in cooler climates. Grows to 6-8′ in 10-15 years.

Foliage yellow in spring, deepening to gold in summer and fall.

Soil: fertile, moist and well drained.

Light: best in full sun. Protect from drying winds.

Pruning: rarely needed. Shear in early spring to shape plant if desired.

Uses: rock garden, accent, patio.

BLUE ITALIAN CYPRESS *Cupressus sempervirens 'Glauca'*

Zones 7 8 9 10
(not southern Florida or Hawaii)

Fast growing narrow columnar tree; grows 2′ or more each year when young. Height to 20-40′ with width of 2-3′. Most dense growth where summers are hot and dry, to reduce growth rate.

Foliage dense, blue green color all year.

Soil: almost any, well drained.

Light: full sun. Young trees require staking until established.

Pruning: rarely needed; inner branches may require tying in to prevent open center.

Uses: specimen or accent, in groups; screen or windbreak, planted at 8-12′.

AMES JUNIPER *Juniperus chinensis 'Ames'*

Zones 4 5 6 7 8 9 10

Pyramidal shrub. Height to 7′ in about 12 years.

Foliage steel blue, color holds all year.

Soil: almost any, well drained; tolerates dry, rocky, acid and alkaline conditions.

Light: best in full sun, will tolerate light shade.

Prunihg: prune and shear as desired.

Uses: specimen or accent, foundation planting.

ARMSTRONG JUNIPER *Juniperus chinensis 'Armstrongii'*

Zones 4 5 6 7 8 9 10

Moderate growth rate, symmetrical and close knit, semi-erect branches. Spreads 4-6′ with height approx. 4′.

Foliage light gray-green, lacy textured. Color holds all year.

Soil: almost any, well drained; tolerated dry, rocky, acid and alkaline conditions.

Light: best in full sun, will tolerate light shade.

Pruning: prune and shear as desired.

Uses: foundation planting, hedge, specimen or accent.

BLAAUW JUNIPER *Juniperus chinensis 'Blaauw'*

Zones 4 5 6 7 8 9 10

Slow growing, semi-erect branches. Grows to height and spread of approx. 4′ in 10-15 years. Eventually spreads to 4-6′ if unpruned, mature height 5-10′.

Foliage deep gray green; color holds all year.

Soil: almost any, well drained; tolerates dry, rocky, acid or alkaline conditions.

Light: best in full sun.

Pruning: easy to shape; clip or shear during growing season.

Uses: specimen or accent, foundation planting.

HETZ JUNIPER *Juniperus chinensis 'Hetzii'*

Zones 4 5 6 7 8 9 10

Very rapid growth, spreading and irregular if unpruned. Arching, semi-erect branches. Spreads 5-10′ in 15 years, matures 10-15′ across. Height 5-10′.

Foliage strong blue-green; color holds all year.

Soil: almost any, well drained: tolerates dry, rocky, acid and alkaline conditions.

Light: best in full sun

Pruning: prune and shear as desired. Fills out quickly

Uses: foundation planting, hedge, specimen or accent, ground cover.

KETELEER JUNIPER *Juniperus chinensis 'Keteleeri'*

Zones 4 5 6 7 8 9 10

Broad pyramid, fast growth, height approx. 10′ in 10-15 years, matures to 20′.

Foliage green; color holds all year.

Soil: almost any, well drained; tolerates dry, rocky, acid and alkaline conditions.

Light: best in full sun.

TOLERATES salt drift from roads.

Pruning: clip or shear during growing season.

Uses: specimen or accent, background, and in large rockeries.

MANEY JUNIPER *Juniperus chinensis 'Maneyi'*

Zones 3 4 5 6 7 8 9 10

Rapid growth, spreading and dense. Semi-erect branches. Spreads 4-8′ in 10-15 years, matures 10-15′ across. Height 3-6′.

Foliage bluish; color holds all year.

Soil: almost any, well drained; tolerates dry, rocky, acid and alkaline conditions.

Light: best in full sun, will tolerate light shade.

Pruning: prune and shear as desired. Bushy growth.

Uses: foundation planting, hedge, specimen or accent.

COLUMNAR CHINESE JUNIPER *Juniperus chinensis 'Mas'*

Zones 4 5 6 7 8 9 10

Dense column, spreads to half height. Reaches 10′ in 10-15 years, matures to approx. 20′.

Foliage gray green; color holds all year.

Soil: almost any, well drained; tolerates dry, rocky, acid and alkaline conditions.

Light: best in full sun.

Pruning: clip or shear during growing season if necessary.

Uses: specimen or accent, background and in large rockeries.

PFITZER JUNIPER *Juniperus chinensis 'Pfitzeriana'*

Zones 4 5 6 7 8 9 10

Very rapid growth, spreading and irregular if unpruned. Arching, semi-erect branches. Spreads 8-12′ in 10-15 years, matures to approx 15′ across. Height 3-8′.

Foliage green with grayish overcast; holds color all year.

Soil: almost any, well drained; tolerates dry, rocky, acid and alkaline conditions.

Light: full sun or light shade.

Pruning: clip or shear during growing season. Prune to shape and to encourage dense growth.

Uses: large rock gardens, foundation planting, ground cover, specimen or massed.

GOLDEN PFITZER JUNIPER *Juniperus chinensis 'Pfitzeriana Aurea'*

Zones 4 5 6 7 8 9 10

Very rapid growth, spreading and irregular if unpruned. Arching, semi-erect branches. Spreads 6-8′ in 10-15 years, matures to approx. 15′ across. Height 3-8′.

Foliage golden yellow when young, color more pronounced when growing on heavy clay soils; matures green with grayish overcast.

Soil: almost any, well drained; tolerates dry, rocky, acid and alkaline conditions.

Light: full sun or light shade.

Pruning: clip or shear during growing season. Prune to shape and to encourage dense growth.

Uses: large rock gardens, foundation planting if color is needed, ground cover, specimen or massed.

BLUE PFITZER JUNIPER *Juniperus chinensis 'Pfitzeriana Glauca'*

Zones 4 5 6 7 8 9 10

Very rapid growth, spreading and irregular if unpruned. Arching, semi-erect branches. Spreads 8-12′ in 10-15 years. Matures to approx. 15′ across.. Height 3-8′.

Foliage strong silvery blue; color holds all year.

Soil: almost any, well drained; tolerates dry, rocky, acid and alkaline conditions.

Light: full sun or light shade.

Pruning: clip or shear during growing season. Prune to shape and to encourage dense growth.

Uses: large rock gardens, foundation plantings, ground cover, specimen or massed.

NICK'S COMPACT PFITZER JUNIPER *Juniperus chinensis 'Pfitzeriana' Nick's Compact*

Zones 4 5 6 7 8 9 10

Slow growth rate with low dense habit. Spreads to 6′ with height only about 12″ in about 10-15 years.

Foliage: Green with grayish overcast; color holds all year.

Soil: Almost any well drained; tolerates dry, rocky, acid & alkaline conditions.

Light: Full sun or light shade.

Pruning: Rarely needed. Clip to shape during growing season.

Uses: Rock gardens, foundation planting, ground cover.

SARGENT JUNIPER *Juniperus chinensis 'Sargentii'*

Zones 3 4 5 6 7 8 9 10

Fairly rapid growth, spreading, branches nearly horizontal. Spreads approx. 10′ in 10-15 years, eventually 10-15′ across. Height 1-2′.

Foliage blue-green, becoming steel blue as plants mature. Color holds all year.

Soil: almost any, well drained; tolerates dry, rocky, acid and alkaline conditions.

Light: best in full sun.

Pruning: rarely needed.

Uses: rock gardens, ground cover and on gentle slopes and banks, patio.

HOLLYWOOD or TWISTED JUNIPER *Juniperus chinensis 'Torulosa'*

Zones 6 7 8 9 10

Narrow upright habit with naturally twisted side branches, slow growing.

Foliage dark green; color holds all year.

Soil: almost any, well drained; tolerates dry, rocky, acid and alkaline conditions.

Light: best in full sun.

Pruning: none needed, but can be sheared.

Uses: rock gardens, specimen or accent.

IRISH JUNIPER *Juniperus communis 'Stricta'*

Zones 3 4 5 6 7 8 9 10

Narrow upright column with moderate growth rate. Height 5-8′ in 10-15 years, matures to approx. 20′. Short lived.

Foliage deep blue green; color holds all year.

Soil: almost any, well drained; tolerates dry, rocky, acid and alkaline conditions.

Light: best in full sun.

Pruning: height can be controlled with light shearing; this encourages dense growth. Branches tend to part with age and snow and may require tying.

Uses: formal specimen or accent.

SPINY GREEK JUNIPER *Juniperus excelsa 'Stricta'*

Zones 6 7 8 9 10

Compact, broad cone shape with slow growth rate. Height 5-8′ in 10-15 years.

Foliage blue green and spiny; color holds all year.

Soil: almost any, well drained; tolerates dry, rocky, acid and alkaline conditions.

Light: best in full sun.

Pruning: light trimming only required.

Uses: specimen or accent.

BLUE HEAVEN or BLUE HAVEN JUNIPER *Juniperus scopulorum 'Blue Heaven' or Blue Haven'*

Zones 3 4 5 6, 7 8 9 10

Symmetrical pyramidal with moderate growth rate. Height 6-8′ in 10 years, matures to 15-20′.

Foliage intense blue color all year.

Soil: almost any, well drained; tolerates dry, rocky, acid and alkaline conditions.

Light: best in full sun.

Pruning: use hand clippers to remove tips of individual long branches. Shear during growing season for formal effect.

Uses: specimen or accent, hedge or screen.

MOFFETT JUNIPER *Juniperus scopulorum 'Moffetti'*

Zones 3 4 5 6 7 8 10

Dense pyramid with moderate growth rate. Height 5-10′ in 10-15 years, matures to approx. 25′.

Foliage intense blue-green with feathery appearance.

Soil: almost any, well drained; tolerates dry, rocky, acid and alkaline conditions.

Light: best in full sun.

Pruning: little shearing required to maintain compact pyramid.

Uses: specimen or accent, background and in large rockeries.

MOONGLO JUNIPER *Juniperus scopulorum 'Moonglo'*

Zones 3 4 5 6 7 8 9 10

Symmetrical pyramid with moderate growth rate. Height 6-8′ in 10-15 years, matures to 15-20′.

Foliage intense blue color; color holds all year.

Soil: almost any, well drained; tolerates dry, rocky, acid and alkaline conditions.

Light: best in full sun.

Pruning: use hand clippers to remove tips of individual long branches. Shear during growing season for formal effect.

Uses: specimen or accent, hedge or screen.

MEYER JUNIPER *Juniperus squamata 'Meyeri'*

Zones 5 6 7 8 9 10

Irregular upright habit, moderate growth rate. Height 3-5′, width 2-4′ in 15-20 years.

Foliage rich blue green when mature, new growth often pinkish. Small black fruits.

Soil: almost any, well drained.

Light: best in full sun, tolerates a little shade.

Pruning not recommended.

Uses: accent, in large rock garden, container plant for patio use.

BURKI JUNIPER *Juniperus virginiana 'Burkii'*

Zones 3 4 5 6 7 8 9 10

Broad pyramid with moderate growth rate. Height 6-9′ in 10-15 years, matures to 15-20′.

Foliage steel blue and dense; turns pink to purple in late fall and winter.

Soil: almost any, well drained; tolerates dry, rocky, acid and alkaline conditions.

Light: best in full sun.

Pruning: shear lightly during growing season.

Uses: specimen or accent.

CANAERT JUNIPER *Juniperus virginiana 'Canaerti'*

Zones 3 4 5 6 7 8 9 10

Compact pyramid, fast growth. Height 10-15′ in 10-15 years, matures to 20-25′.

Foliage dark green with tufted appearance.

Color holds all year. Heavy crops of small pale blue berries.

Soil: almost any, well drained; tolerates dry, rocky, acid and alkaline conditions.

Light: best in full sun.

Pruning: shear or clip during growing season.

Uses: specimen or accent, hedge or screen.

DUNDEE JUNIPER *Juniperus virginiana 'Hillii'*

Zones 4 5 6 7 8 9 10

Narrowly conical with moderate growth rate. Height 8-10′ in 10-15 years, matures to approx. 15′. Grows faster in open position.

Foliage gray-green, turns purplish during winter.

Soil: almost any, well drained; tolerates dry, rocky, acid and alkaline conditions.

Light: best in full sun.

Pruning: shear or clip during growing season.

Uses: specimen or accent, hedge or screen.

SKYROCKET JUNIPER *Juniperus virginiana 'Skyrocket'*

Zones 5 6 7 8 9 10

Narrow conical tree with moderate growth rate. Height 6-8′ in 10-15 years, matures to 15-20′.

Foliage blue gray, becoming duller in winter.

Soil: almost any, well drained; tolerates rocky, dry, acid and alkaline conditions.

Light: best in full sun.

Pruning: use hand clipper to remove tips of individual long branches. Shear during growing season for formal effect.

Uses: specimen or group.

NORWAY SPRUCE *Picea abies*

Zones 3 4 5 6 7,
plus northwest & east areas in 8,
and north in zone 10

Fast growing tree, symmetrical when young. Height 10-15′ in about 10 years, matures to 80-120′

Needles dark green; color holds all year.

Soil: moist and well drained; water well in light, sandy conditions.

Light: full sun

TOLERANT of wind

Pruning: rarely needed; tolerates trimming.

Uses: background, windbreak, screen or hedge specimen or accent tree

BIRDS NEST SPRUCE *Picea abies 'Nidiformis'*

Zones 3 4 5 6 7,
plus northwest & east areas in 8,
and north in zone 10

Very slow growing and spreading to 2-3′ across with height of 1-1½′ in 10 years. Matures to 6-10′ wide and 2-5′ high.

Needles dark green; color holds all year.

Soil: moist and well drained: water well in light sandy conditions.

Light: full sun or light shade

Pruning: none needed.

Uses: rock gardens, foundation planting, ground cover.

ENGELMANN SPRUCE *Picea engelmannii 'Glauca'*

Zones 3 4 5 6 7 8
and north in zone 10

Slow growing conical tree. Height 6-10′ in about 10 years, matures to 60-80′.

Needles blue green, densely packed; color holds all year. Conspicuous brown buds in winter.

Soil: moist and well drained; water well in light sandy conditions.

Light: full sun

TOLERATES wind

Pruning: rarely needed.

Uses: specimen or accent tree, background.

WHITE SPRUCE *Picea glauca*

Zones 3 4 5 6 7 8
and north in zone 10

Slow growing conical tree with short branches. Mature height 20-30′

Needles bluish dark green; color holds all year.

Soil: moist and well drained.

Light: full sun.

TOLERATES wind.

Pruning: rarely needed; tolerates trimming.

Uses: background, windbreak, screen or hedge specimen or accent tree.

DWARF ALBERTA SPRUCE *Picea glauca 'Conica'*

Zones 3 4 5 6 7 8
and north in zone 10

Symmetrical, broadly conical dwarf tree. Height 7′ in 25-30 years.

Needles short and tightly placed, light green; color holds all year.

Soil: moist and well drained; water well in light sandy conditions.

Light: full sun

TOLERATES wind

Pruning: shear lightly to retain formal shape.

Uses: small accent tree

BLACK HILLS SPRUCE *Picea glauca 'Densata'*

Zones 3 4 5,
plus northwest areas in 6 7 & 8,
and north in zone 10.

Compact and bushy; sturdy, slow growing tree. Height 8-12′ in about 10 years, matures 30-50′.

Needles short, bluish dark green; color holds all year.

Soil: almost any, moist and well drained.

Light: full sun.

TOLERATES wind, heat, drought (once established), cold, and overcrowding.

Pruning: tolerates trimming.

Uses: background, windbreak, screen or hedge specimen or accent tree

BLACK SPRUCE *Picea mariana*

Zones 3 4 5 6 7 8
and north in zone 10

Slow growing broadly conical tree. Height reaches 10-15′ in about 15 years, matures to 20-30′.

Needles short, dull or bluish dark green; color holds all year.

Soil: moist and well drained; water well in light sandy conditions.

Light: full sun.

TOLERATES wind.

Pruning: rarely needed.

Uses: specimen or accent tree.

COLORADO SPRUCE *Picea pungens*

Zones 3 4 5 6 7 8
and north in zone 10

Broadly conical tree, rigid branches with stiff needles. Height 5-10′ in about 10 years, matures to 80-120′.

Needles light green; color holds all year.

Soil: moist and well drained.

Light: full sun.

TOLERATES wind.

Pruning: rarely needed; trim in July if necessary.

Uses: specimen or accent tree; needs fairly large space.

COLORADO BLUE SPRUCE *Picea pungens 'Glauca'*

Zones 3 4 5 6 7 8
and north in zone 10

Broadly conical tree, rigid branches with stiff needles. Height 6-10′ in about 5 years, matures to 60-80′.

Needles rich glistening blue; color holds all year.

Soil: moist and well drained.

Light: full sun.

TOLERATES wind.

Pruning: rarely needed; trim in July if necessary.

Uses: specimen or accent tree; needs fairly large space. Roadside tree.

SWISS STONE PINE *Picea pungens 'Moerheim'*

Zones 3 4 5 6 7 8
and north in zone 10

Broadly conical to tapering tree, rigid branches with stiff needles. Slow growing, height 4-6′ in about 10 years, matures to 25-35′.

Needles strong blue; color holds all year.

Soil: moist and well drained.

Light: full sun.

TOLERATES wind.

Pruning: rarely needed.

Uses: specimen or accent tree; needs fairly large space.

SLASH OR CUBAN PINE *Pinus caribaea*

Zones 9 10

Fast growing when young, adds 2-3′ to height each year. Matures to 70-100′ as timber tree.

Needles glossy green, in groups of three, 8-12″ long.

Soil: tolerates both moist and dry conditions.

Light: full sun.

Pruning: pinch back new young growth in spring or cut in mid June.

Uses: specimen or accent, fast growing screen.

SWISS STONE PINE *Pinus cembra*

Zones 3 4 5 6 7 8,
plus north areas in zone 10

Compact, tight pyramid with slow growth rate. Height reaches 25′ in 25-30 years, spreads 10-15′.

Needles dark blue-green, densely packed, in groups of five, 2½-4½″ long.

Soil: almost any, well drained; tolerates poor, rocky and sandy conditions.

Light: full sun or light shade.

TOLERATES coastal climate.

Pruning: rarely needed.

Uses: good in small gardens, specimen or accent, good for seaside conditions.

ALEPPO PINE *Pinus halepensis*

Zones 7 8 9 10

Fast growing broad open tree adding 3-5′ to height each year when young. Height 30-40′ in 15-20 years.

Needles light green in pairs, 2½-4″ long.

Cones 3″ long.

Soil: almost any; tolerates poor, rocky, sandy, moist or dry conditions.

Light: full sun.

TOLERATES heat, drought, wind, coastal and desert climates.

Pruning: rarely needed.

Uses: large specimen or group, screen, especially good for seaside or desert conditions.

MUGO PINE
or SWISS MOUNTAIN PINE *Pinus mugo*

Zones 3 4 5 6 7 8,
plus north in zone 10

Slow growing spreading tree; width 8-10′ and height 3-10′ in 15-20 years.

Needles bright to dark green, in pairs.

Soil: almost any, well drained; tolerates poor, rocky and sandy conditions.

Light: full sun or light shade.

Pruning: pinch or cut back new growth in spring or mid-June to encourage spread.

Uses: foundation planting, specimen or accent, rock garden or border, low hedge, ground cover, patio, Bonsai.

AUSTRIAN or AUSTRIAN BLACK PINE *Pinus nigra*

Zones 4 5 6 7 8, plus north in zone 10

Symmetrical when young, loses lower branches with age. Rapid growth to height 10-15′ in 15 years; matures to 50-100′.

Needles in pairs, very dark green contrasting with yellow-green of new growth each spring.

Soil: almost any; tolerates poor, rocky, sandy, alkaline and clay conditions.

Light: full sun.

TOLERATES urban polution and coastal climates.

Uses: screens and windbreaks, specimen or accent, roadside, where salt spray drifts.

ITALIAN STONE PINE *Pinus pinea*

Zones 7 8 9 10
(not in southern Florida)

Flat topped or umbrella shaped tree with moderate growth rate. Branches cluster at top of trunk, making asymmetrical head if not trimmed. Height 25-30′ in 25-30 years.

Bark red-gray color; deeply grooved.

Needles bright green, 4-8″ long, in pairs.

Cones 5″ long. Seeds edible.

Soil: almost any, well drained; tolerates poor, rocky and sandy conditions.

Light: full sun to light shade.

Pruning: to shape growth.

Uses: specimen or accent, street tree.

WESTERN YELLOW PINE *Pinus ponderosa*

Zones 6 7 8

Narrowly conical tree, fast growing when young. Height 12-18′ in 10 years, matures 75-100′.

Needles long and dark green, in groups of three densely packed. Reddish-brown scaly bark.

Soil: almost any, well-drained; tolerates poor, rocky, sandy and alkaline conditions.

Light: full sun.

Pruning: not necessary, unless dwarfing is desired.

Uses: large specimen, screen.

MONTEREY PINE *Pinus radiata*

Zones 7 8 9 10

Fast growing tree when young, matures to 60′. Naturally well shaped tree.

Needles bright green in threes, 4-6″ long.

Cones 3-5½″ long.

Soil: almost any; tolerates poor, rocky, sandy, moist or dry conditions.

Light: full sun.

TOLERATES heat, drought, wind, coastal climate.

Pruning: rarely needed.

Uses: specimen or group, screen, especially good for seaside conditions.

NORWAY or RED PINE *Pinus resinosa*

Zones 3 4 5 6 7 8
and north in zone 10

Native to Norway, Maine. Broad pyramid, symmetrical and fast growing when young. Height approx. 50′ in 25-30 years. Best in cool climates.

Needles in pairs, long and flexible, dark green.

Soil: almost any well drained; tolerates poor, rocky, sandy, moist or dry conditions.

Light: full sun.

Uses: specimen or accent.

EASTERN WHITE PINE *Pinus strobus*

Zones 3 4 5 6 7 8
and north in zone 10

Fast growing pyramid when young, loses lower branches with age. Height 18-25′ in 15 years, matures to 75-130′.

Needles in groups of five, soft and blue-green.

Soil: almost any, well drained; tolerates poor, rocky, light and sandy conditions.

Light: full sun though needs light shade in extreme heat.

Uses: specimen or accent, screen or windbreak.

SCOTCH PINE *Pinus sylvestris*

Zones 3 4 5 6 7 8
and north in zone 10

Fast growing pyramid when young, matures irregular or flat topped. Height 12-18′ in 15 years, matures to 40-80′.

Needles in pairs, blue-green.

Soil: almost any well drained; tolerates poor, light, sandy, moist or dry conditions.

Light: full sun.

Uses: specimen or accent.

BERCKMANN'S GOLDEN ARBORVITAE *Platycladus orientalis (Thuja orientalis) 'Berckmannii'*

Zones 6 7 8 9 10

Compact oval or globe-shaped tree. Height to 5′ in 8-10 years if unpruned.

Foliage golden in spring, turning light green in summer and bronze in winter. Winter damage may occur in northern areas of recommended zones.

Soil: almost any, moist and well drained.

Light: full sun or shade.

Pruning: needs little to promote dense growth; easy to control height.

Uses: specimen or accent, low dense hedge, space 18-36″.

SOUTHERN or JAPANESE YEW *Podocarpus macrophyllus and Podocarpus macrophyllus var. maki*

Zones 7 8 9 10

Shrub or tree with horizontal, slightly pendulous branches. Height 5-8′ in 10-15 years; *P. macrophyllus* matures to 40-50′, *P. m. maki* remains smaller.

Needles 3-4″ long, flexible and dark green.

Soil: moist and well drained.

Light: full sun or light shade; shade needed where summers are hot.

Pruning: clip individual branches to shape.

Uses: patio, container plant, specimen or accent, good for small gardens (*P. m. maki*).

NOTE: do not plant against sunny wall since the reflected heat will damage Podocarpus.

DOUGLAS FIR *Pseudotsuga menziesii*

Zones 3 4 5 6 7 8
and north in zone 10

Very fast growing symmetrical pyramid. Adds 1-2′ to height every year when young, matures to 100′ or more.

Needles aromatic, blue or blue-green, soft to touch; color holds all year. 4″ cones hang from branches of older trees.

Soil: moist and well drained.

Light: best in full sun.

Prune: in early spring before new growth starts.

Uses: specimen or accent tree; best in fairly large area screen.

JAPANESE YEW *Taxus cuspidata*

Zones 5 6 7, plus north and west in 8, west in 4, and north in zone 10

Spreading shrub with rapid growth rate; adds 12-18″ to height and spread each year. If unpruned, reaches height of 10-12′ and 10-15′ width in about 15 years.

Needles dark green; protect from winds to minimize winter burn.

Soil: acid to slightly acid, fertile, moist and well drained.

Light: full sun to deep shade.

Pruning: clip or shear in early summer to thicken and shape.

Uses: foundation planting, specimen or accent, hedge.

UPRIGHT JAPANESE YEW *Taxus cuspidata 'Capitata'*

Zones 5 6 7,
plus north and west in 8,
west in 4, and north in zone 10

Broad conical shrub with fairly rapid growth rate; adds 10-12″ to height each year if unpruned. Easy to maintain as hedge or specimen. Clear off snow accumulations to prevent hollow center.

Needles dark green; protect from winds to minimize winter burn.

Soil: acid to slightly acid, fertile, moist and well drained.

Light: full sun to deep shade.

Pruning: clip or shear in early summer to thicken and shape.

Uses: foundation planting, specimen or accent, hedge or screen; good to height of 12′.

DENSE SPREADING JAPANESE YEW *Taxus cuspidata 'Densa'*

Zones 5 6 7,
plus north and west in 8,
west in 4, and north in zone 10

Slow growing compact shrub. If unpruned, spreads to 5-8′ across with height of 3-4′ in 10-15 years.

Needles rich dark green; protect from winds to minimize winter burn.

Soil: slightly acid, fertile, moist and well drained.

Light: full sun to deep shade.

Pruning: clip or shear in early summer to thicken and shape.

Uses: foundation planting, specimen or accent, low hedge or edging.

DWARF JAPANESE YEW *Taxus cuspidata 'Nana'*

Zones 5 6 7,
plus north and west in 8,
west in 4, and north in zone 10

Slow growing shrub, adds 6-12″ to height and spread each year. If unpruned, will reach height of 4-6′ with 8-10′ spread in 15-20 years.

Needles dark green; protect from winds to minimize winter burn. Female plants bear abundant red fruit.

Soil: slightly acid, fertile, moist and well drained.

Light: partial sun to deep shade.

Pruning: necessary to prevent irregular growth.

Uses: foundation planting, specimen or accent, low hedge or edging.

BROWN'S YEW *Taxus ×media 'Brownii'*

Zones 5 6 7,
plus north and west in 8,
west in 4, and north in zone 10

Compact shrub, semi-erect branches, moderate growth rate. If unpruned, spreads 6-10′ with height of 5-8′ in 10-15 years.

Needles dark green; resists heat and drought damage, including winter burn.

Soil: slightly acid, fertile and moist, well drained.

Light: full sun to deep shade.

Pruning: easily pruned to globe shape.

Uses: foundation planting, specimen or accent, low hedging or edging.

HICKS UPRIGHT YEW *Taxus ×media 'Hicksii'*

Zones 5 6 7,
plus north and west in 8,
west in 4, and north in zone 10

Upright habit with moderate growth rate. Height 6-10′, width 3-5′ in 10-15 years if unpruned.

Needles very dark green; protect from winds to minimize winter burn.

Soil: slightly acid, fertile and moist, well drained.

Light: full sun to deep shade.

Pruning: easily shaped by shearing or clipping.

Uses: foundation planting, specimen or accent, screen or hedge to 6′.

SEBIAN YEW *Taxus ×media 'Sebian'*

Zones 5 6 7,
plus north and west in 8,
west in 4, and north in zone 10

Slow growing compact shrub. If unpruned, spreads to 5-8′ across with height of 3-4′ in 10-15 years.

Needles intense dark green; resists winter burn.

Soil: slightly acid, fertile and moist, well drained.

Light: full sun to deep shade.

Pruning: clip or shear in early summer to thicken and shape.

Uses: foundation planting, specimen or accent, low hedge or edging.

TAUNTON SPREADING YEW *Taxus ×media 'Tauntonii'*

Zones 5 6 and 7,
plus north and west in 8,
west in 4, and north in zone 10

Free branching shrub with semi-erect branches. Spreads 6-10′ with height of 5-8′ in 10-15 years if unpruned.

Needles dark green; resists winter damage.

Soil: slightly acid, fertile and moist, well drained.

Light: full sun to shade.

Pruning: little required to promote thick growth; contain size and shape by shearing or clipping.

Uses: foundation planting, specimen or accent, low hedge or edging.

WARD'S YEW *Taxus ×media 'Wardii'*

Zones 5 6 and 7,
plus north and west in 8,
west in 4, and north in zone 10

Low spreading shrub; spreads to 5-10′ across with height of 3-5′ in 15 to 20 years if unpruned.

Needles dark green, with new growth yellow-green color. Resists winter damage.

Soil: slightly acid, fertile and moist, well drained.

Light: full sun to deep shade.

Pruning: little required to maintain low form.

Uses: foundation planting, specimen or accent, low hedge or edging.

GLOBE ARBORVITAE *Thuja occidentalis 'Globosa'*

Zones 3 4 5 6 7 8
and north in zone 10

Slow growing globose tree. Rarely exceeds 3-5′ high with spread of 4-6′.

Foliage rich dark green; color holds to late winter when some bronzing will occur in colder zones.

Soil: almost any, moist and well drained.

Light: full sun or shade.

Pruning: none needed to maintain globose shape.

Uses: specimen or accent, foundation planting, patio.

DARK GREEN or WINTERGREEN ARBORVITAE *Thuja occidentalis 'Nigra'*

Zones 3 4 5 6 7
and north in zone 10

Broadly conical fast growing tree. Height 15-20′ in 15 years if unpruned.

Foliage rich green; color holds all year.

Soil: almost any moist and well drained.

Light: full sun or shade.

Pruning: easy to control height by clipping and shearing.

Uses: hedge or screen, spaced at 18-36″ specimen or accent.

PYRAMIDAL ARBORVITAE *Thuja occidentalis 'Pyramidalis'*

Zones 3 4 5 6 7 8
and north in zone 10

Narrow pyramid with moderate to fast growth rate. Height 4-8′ in 10 years, matures 15-20′.

Foliage strong mid-green; tends to bronze in mid to late winter.

Soil: almost any, moist and well drained.

Light: full sun to deep shade.

Pruning: easy to control height by clipping and shearing.

Uses: good in east or north locations, specimen or accent hedge or screen, spaced at 18-36″.

RHEINGOLD ARBORVITAE *Thuja occidentalis 'Rheingold'*

Zones 3 4 5 6 7 8
and north in zone 10

Broadly conical, fast growing tree. Height to 5-8′ in 10-15 years, matures 10-15′. Juvenile form is rounded compact bush.

Foliage bright yellow in spring, turning gold in summer and bronze in winter.

Soil: almost any, moist and well drained.

Light: full sun or shade.

Pruning: easy to maintain by clipping and shearing.

Uses: specimen or accent, hedge or screen, spaced at 18-36″.

TECHNY, or MISSION ARBORVITAE *Thuja occidentalis 'Techny'*

Zones 3 4 5 6 7 8
and north in zone 10

Fast growing conical tree. Height to approx. 8′ in 10 years, matures to 15-30′ if unpruned.

Foliage very dark green; color holds all year.

Soil: almost any; tolerates moist conditions.

Light: full sun or light shade.

Pruning: easy to control height by clipping and shearing.

Uses: hedge or screen, spaced at 18-36″ specimen or accent.

SIBERIAN or WARE ARBORVITAE *Thuja occidentalis 'Wareana'*

Zones 3 4 5 6 7 8
and north in zone 10

Slow growing, dense, broadly conical tree. Height 3-6′ in 10 years, matures to 12-18′ if unpruned.

Foliage light yellow-green when young, dark green in summer; color holds well all year.

Soil: almost any, moist and well drained.

Light: full sun or light shade.

Pruning: needs little to promote dense growth; easy to control height.

Uses: specimen or accent, foundation planting, hedge or screen, spaced at 18-36″.

CANADIAN HEMLOCK *Tsuga canadensis*

Zones 4 5 6 7,
plus northwest and northeast areas in 8,
and north in zone 10

Broad pyramid with arching, drooping branches; fast growth rate. Height 12-16′ in 10 years, matures to 80-100′ if left unpruned.

Needles soft and dark green with whitish undersides; color holds all year.

Soil: moist, deep, well drained; prefers acid conditions though will tolerate some lime if peatmoss, leafmold or compost is added.

Light: full sun or shade. Shelter from winds at all times.

Pruning: shear for formal effect; very tolerant of clipping.

Uses: good specimen for dark corners, north side of building hedge or screen.

Deciduous Shrubs

AMUR MAPLE *Acer ginnala*

Zones 3 4 5 6 7 8
and north area in zone 10

Shrubby multistemmed bush or small single stem tree. Moderate growth rate. Height 20′ in 10-12 years, matures 25-35′; spread equals height.

Flowers in early spring while leaves unfurl. Tiny white blossoms followed by bright red winged fruit, in midsummer.

Leaves smoothly lobed, 2-3½″ long; pale green in spring, turning dark green in summer.

FALL COLOR: yellow, orange, scarlet.

Soil: almost any, tolerates wet or dry conditions.

Light: full sun to light shade.

Prune in late summer and fall, to shape and train as desired.

Uses: screen or hedge, shrub border, specimen or accent.

REDLEAF JAPANESE MAPLE *Acer palmatum 'Atropurpureum'*

Zones 4 5 6 7 10,
plus east areas in zone 8.

Small deciduous tree or shrub. Slow growing. Height approx. 15′ in 10-15 years, matures 15-20′; spread equals height. Grows to larger size in south.

Leaves deeply cut, red purple color all season. Color intensifies in fall.

Soil: moist and well drained; tolerates lime.

Light: partially shaded position with protection from cold winds.

Pruning: rarely needed.

Uses: accent or specimen, patio tree.

JAPANESE BARBERRY *Berberis thunbergii*

Zones 4 5 6 7 8 9 10

Deciduous shrub with arching branches. Unpruned height reaches 6-8′; spreads 5-8′. Single spines line branches.

Flowers in spring, profuse light-yellow blossoms along stems. Red fruit follow flowers, persisting into winter if not taken by birds.

Leaves ½-1½″ long, yellow green when young; mature green.
FALL COLOR: orange red.

Soil: almost any, well drained.

Light: best in full sun.

TOLERATES heat and drought.

IMMUNE to wheat rust disease.

Prune to shape; shear several times each season to maintain dense, formal appearance. Specimen plants rarely need pruning.

Uses: low hedge, specimen or accent, foundation.

GOLDEN BARBERRY *Berberis thunbergii 'Aurea'*

Zones 4 5 6 7 8 9 10

Deciduous shrub with arching branches. Unpruned height reaches 8′; spread 5-8′. Single spines line branches.

Flowers in spring, profuse light-yellow blossoms along stems. Red fruit follow flowers, persisting into winter if not taken by birds.

Leaves ½-1½″ long, golden yellow color all summer. FALL COLOR: yellow orange.

Soil: almost any, well drained.

Light: best in full sun.

TOLERATES heat and drought.

IMMUNE to wheat rust disease.

Prune to shape; shear several times each season to maintain dense, formal appearance. Specimen plants rarely need pruning.

Uses: low hedge, specimen or accent, foundation.

RED BARBERRY *Berberis thunbergii 'Atropurpurea'*

Zones 4 5 6 7 8 9 10

Erect small shrub forming a dense column. Height to 4′ in hedges. Single spines line branches.

Flowers in spring: yellow blossoms along stems, followed by persistent red berries.

Leaves: shiny dark wine-red color holds all summer, intensifies in fall.

Soil: almost any, well drained.

Light: best in full sun.

TOLERATES heat and drought.

IMMUNE to wheat rust disease.

Prune to shape; shear several times each season to maintain dense, formal appearance.

Uses: hedge or barrier.

DWARF RED OR CRIMSON PIGMY BARBERRY *Berberis thunbergii 'Atropurpurea Nana'*

Zones 4 5 6 7 8 9 10

Dwarf deciduous shrub, grows to height of 6-24″ with spread of 12-36″.

Leaves ½″ long, brilliant red all summer.
FALL COLOR: intensified brilliant red, especially if planted in full sun.

Soil: almost any, well drained.

Light: best in full sun.

TOLERATES heat and drought.

IMMUNE to wheat rust.

Pruning rarely required; shear lightly to shape hedge.

Uses: edging plant, low hedge, front of borders.

SIBERIAN PEA TREE *Caragana arborescens*

Zones 3 4 5 6 7 8 9 (not southern Florida), and 10.

Dense growing shrub with stiff upright branches. Height 10-15′; spreads to 6-8′. Young twigs characteristic yellow green color and winged bark.

Flowers in late spring. Fragrant yellow pea-shaped blossoms cluster along branches. Small pods form in summer.

Leaves pinnate and pea-like, bright green color.

Soil: prefers light and sandy conditions, well drained.

Light: full sun.

TOLERATES wind, drought, heat and cold.

Prune right after flowering, reducing previous year's growth to two thirds its length.

Uses: hedge or windbreak: plant 18-24″ apart. Promote dense growth by cutting to ground in first spring.

JAPANESE FLOWERING QUINCE *Chaenomeles speciosa*

Zones 4 5 6 7 8 9 10

Upright and spreading shrub with stiff spiny branches. Unpruned mature height about 8-10′; spread equal to or greater than height. Branches cut in midwinter will flower indoors.

Flowers in early spring before leaves unfurl. Continues to flower sparingly right through summer. Single blossoms in reds, pinks, or white, line the branches.

Fruit ripens in fall; greenish yellow and edible (can be used to make jelly).

Leaves bronze-red in spring, turning glossy dark green in summer.
FALL COLOR: red-bronze.

Soil: almost any.

Light: full sun.

Prune after flowering, to shape hedge or specimen.

Uses: shrub borders, foundation planting, hedge, espalier, barrier, specimen.

VARIEGATED or SILVER EDGED DOGWOOD *Cornus alba 'Argenteo-marginata'*

Zones 3 4 5 6 7 8
and north in zone 10

Upright deciduous shrub, unpruned height to 7-9′. Best pruned to encourage new growth for best stem color and greater vigor. Stem color bright red.

Flowers in early June, whitish and fragrant blossoms in 2″ clusters.

Fruit: white, ripens in late summer.

Leaves: dark green with white edges throughout summer.

Soil: almost any, moist and well drained; thrives in damp places, especially near water.

Light: best in full sun.

Prune in early spring before new growth starts. Cut away oldest stems right to ground level, to encourage succession to new shoots (about one third are removed each year).

Uses: specimen or accent, singly or in groups for bark color in winter and leaf color in summer.

SMOKE TREE *Cotinus coggygria*

Zones 4 5 6 7 8
and north in zone 10

Rounded deciduous shrub. Height 10-12′ in 5-8 years, matures to 25′ if unpruned. Spreads 10-15′.

Flowers in early summer on new season's growth. Sterile flower parts extend and become covered with hairs, giving pinkish smoky appearance by late summer.

Leaves rounded or oval, dark green.

FALL COLOR: yellow to brilliant orange.

Soil: almost any, well drained.

Light: full sun.

Pruning rarely required. Remove straggly growth and shape in early spring before new growth starts.

Uses: specimen.

ROYAL PURPLE SMOKE TREE *Cotinus coggygria 'Royal Purple'*

Zones 4 5 6 7 8
and north in zone 10

Rounded deciduous shrub. Height 10-12′ in 5-8 years, matures to 15′ if unpruned. Spreads 10-15′.

Flowers in early summer on new season's growth. Sterile flower parts extend and become covered with hairs, giving bronze colored smoky appearance by late summer.

Leaves rounded or oval, purple color all summer.

FALL COLOR: yellow-orange.

Soil: almost any, well drained.

Light: full sun.

Pruning rarely required. Remove straggly growth and shape in early spring before new growth starts.

Uses: specimen.

PEKING COTONEASTER *Cotoneaster acutifolius*

Zones 5 6 7 8

Medium sized well branched deciduous or semi-evergreen shrub. Height to 12′ if unpruned.

Flowers in spring, pale pink.

Fruit small and black.

Leaves dark green, 1-2½″ long, color held through fall.

Soil: almost any, well drained.

Light: best in full sun.

Pruning rarely needed. Shear lightly in hedge.

Uses: hedges up to 6′ tall; plant at 18-30″ screen.

CRANBERRY COTONEASTER *Cotoneaster apiculatus*

Zones 4 5 6 7 8

Medium sized mounded spreading deciduous shrub; height 2-4′ and spread 5-10′. Semi-evergreen in milder climates.

Flowers in spring, small pinkish-white blossoms.

Fruit persist through winter, glowing red berries that look like cranberries (though tasteless). Fruit ripens in summer.

Leaves ½-1″ long, glossy green with gray undersides.

FALL COLOR (picture): Deep crimson, best color in full sun.

Soil: almost any, well drained.

Light: best in full sun.

Pruning rarely needed.

Uses: low hedge, foundation planting, ground cover.

MANY-FLOWERED COTONEASTER *Cotoneaster multiflorus*

Zones 5 6 7 8

Deciduous shrub with long arching stems. Height to 6-8′; spreading plant that needs space to develop.

Flowers in spring; ½″ white blossoms in clusters; musty odor.

Fruit: large red berries.

Leaves: gray green in summer, ¾-2″ long.

FALL COLOR: reds and oranges.

Soil: almost any, well drained.

Light: best in full sun.

Pruning: rarely needed.

Uses: trained over fence for hedge or screen against wall, espaliered.

WARMINSTER BROOM *Cytisus ×praecox*

Zones 5 6 7 8 9 10

Fast growing deciduous shrub, slender upright branches which curve outwards. Grows to 5-6′ height and spread. Branches green. Hard to transplant large specimens. Short lived on lime soils.

Flowers on previous season's growth. Abundant pea-shaped flowers, fragrant and light yellow, before leaves unfurl. Dry flat pods ripen in late summer, rattle in wind and explode noisily as they dry.

Soil: well drained, tolerates lime.

Light: full sun.

TOLERATES wind, coastal climate, sandy soils.

Prune right after flowering, trimming off two-thirds of the previous season's growth.

Uses: group, accent, foundation planting, hedge.

SLENDER DEUTZIA *Deutzia gracilis*

Zones 4 5 6 7 8 9
(not southern Florida) and 10.

Upright deciduous shrub with slender arching branches. Grows 2-4′ tall with spread of 4-6′.

Flowers in late spring or early summer, on last year's growth. Abundant snow-white clusters cover the young leaves. Small brown seed pods follow flowers.

Soil: almost any, well drained. Protect from north winds.

Light: full sun or light shade.

Prune after flowering to encourage new growth for next year's flowers. Tips of shoots may be killed in winter: remove dead wood in spring.

Uses: low informal hedge, plant 2-2¼′ apart, in group planting, specimen or accent, patio shrub.

SHOWY DEUTZIA *Deutzia × magnifica*

Zones 5 6 7 8 9 (not southern Florida) and 10

Upright deciduous shrub with slender arching stems. Grows 5-7′ tall with spread of 5-8′.

Flowers in late spring or early summer on last year's growth. Abundant double white blossoms in dense clusters. Followed by small brown seed pods.

Soil: almost any, well drained. Protect from north winds.

Light: full sun or light shade.

Prune after flowering to encourage new growth for next year's flowers. Tips may be killed back in winter; remove dead wood in spring.

Uses: specimen or accent shrub; informal hedge, plant 2-2½′ apart; in shrub border.

BURNING BUSH *Euonymus alata 'Compacta'*

Zones 3 4 5 6 7 8 and north in zone 10

An excellent, densely branched, compact deciduous shrub of slow growth. Easy culture, rounded in outline. Matures to 5-6′ high. Branches have bark of corky-like ridges.

Flowers in late spring, small and delicate, yellow-green color.

Leaves 1½-2½″ long, mid green all summer.

FALL COLOR: Unbelievable brilliance of red foliage, more intense in full sun.

Soil: almost any, well drained.

Light: full sun or light shade.

Pruning: rarely needed.

Uses: accent, back of border, foundation planting, hedge.

BURNING BUSH *Euonymus alata 'Monstrosa'*

Zones 3 4 5 6 7 8 and north in zone 10

An excellent, densely branched, deciduous shrub of slow growth. Easy culture, rounded in outline. Matures to 8-12′ high. Branches have bark of corky-like ridges.

Flowers in late spring, small and delicate, yellow-green color.

Leaves 1½-2½″ long, mid green all summer.

FALL COLOR: Unbelievable brilliance of red foliage, more intense in full sun.

Soil: almost any, well drained.

Light: full sun or light shade.

Pruning: rarely needed.

Uses: accent, back of border, foundation planting, hedge.

EUROPEAN SPINDLE TREE *Euonymus europaea*

Zones 3 4 5 6 7 8
and north in zone 10

Vigorous deciduous shrub or small tree, upright and spreading habit. Height to about 10′.

Fruit: large clusters of pink or deep rose fruits with orange seeds.

Leaves 1½-3″ long, dark green.

FALL COLOR: dark red, more intense in full sun.

Soil: almost any, well drained.

Light: full sun or light shade.

Pruning rarely needed.

Uses: specimen or accent, group planting, hedge or screen.

FORSYTHIA *Forsythia ×intermedia*

Zones 4 5 6 7 8 9
and north in zone 10

Vigorous deciduous shrub with long arching branches. Height of freestanding plant will reach 8′. Very easy to establish. Severe winter may kill top portions; quickly grows from lower stems. Branches cut in midwinter will flower indoors.

Flowers profusely in early spring before leaves unfurl. Shade of yellow, lemon to golden.

Soil: almost any, well drained.

Light: full sun or light shade.

TOLERATES urban pollution.

Prune right after flowering.

Uses: hedge, against wall, screen, espalier, free standing specimen or accent.

ALTHEA, ROSE OF SHARON *Hibiscus syriacus*

Zones 4 5 6 7 8 9
(not southern Florida) and 10

Fast growing upright shrub, spreading broadly as plant matures. Branches pliable and easily trained. Freestanding height reaches 6-12′.

Flowers on this year's growth in late summer and fall. Single blossoms in white, pinks, purples, or blues.

Leaves open in late spring; green color.

Soil: moist and well drained; add peatmoss, leafmold or compost.

Light: best in full sun. Protect from winds especially when flowers are open.

Prune in spring to remove winter killed shoot tips. Shorten over-long shoots after flowering.

Uses: in borders, against wall, trellis, screen, accent or in group planting, patio plant.

ALTHEA, ROSE OF SHARON *Hibiscus syriacus*

Zones 4 5 6 7 8 9 (not southern Florida) and 10

Fast growing upright shrub, sometimes treeform, spreading broadly as plant matures. Branches pliable and easily trained. Freestanding height reaches 6-12′.

Flowers on this year's growth in late summer and fall. Single blossoms in white, pinks, purples, or blues.

Leaves open in late spring; green color.

Soil: moist and well drained; add peatmoss, leafmold or compost.

Light: best in full sun. Protect from winds especially when flowers are open.

Prune in spring to remove winter killed shoot tips. Shorten over-long shoots after flowering.

Uses: in borders, against wall, trellis, screen, accent or in group planting, patio plant.

ALTHEA, ROSE OF SHARON *Hibiscus syriacus*

Zones 4 5 6 7 8 9 (not southern Florida) and 10

Fast growing upright shrub, sometimes treeform, spreading broadly as plant matures. Branches pliable and easily trained. Freestanding height reaches 6-12′.

Flowers on this year's growth in late summer and fall. Single blossoms in white, pinks, purples, or blues.

Leaves open in late spring; green color.

Soil: moist and well drained; add peatmoss, leafmold or compost.

Light: best in full sun. Protect from winds especially when flowers are open.

Prune in spring to remove winter killed shoot tips. Shorten over-long shoots after flowering.

Uses: in borders, against wall, trellis, screen, accent or in group planting, patio plant.

HILLS OF SNOW HYDRANGEA *Hydrangea arborescens 'Grandiflora'* Zones 4 5 6 7 8 9 10

Loose spreading deciduous shrub, growing to height 3-4′.

Flowers on this year's growth, July through September; pure white rounded clusters. Stems pulled down with the weight of the flower heads.

Leaves 3-6″ long, bright green ovals.

FALL COLOR: bronze to wine red.

Soil: moist and well drained; add peatmoss, leafmold or compost. Apply mulch each spring after soil has warmed, to retain moisture and feed the surface roots.

Light: best in partial shade; tolerates afternoon sun.

Prune back previous season's growth by about half, in late winter. For larger, fewer flowers, cut back to 2 good buds. Remove dead flower heads in fall or early spring.

Uses: shrub border, accent, informal hedge planted 2-3′ apart.

ANNABELLE HYDRANGEA *Hydrangea arbeorescens 'Grandiflora-Annabelle'*

Zones 4 5 6 7 8 9 10

Loose spreading deciduous shrub, growing to height of 3-4′.

Flowers on this year's growth, July through September; snowy white rounded clusters 10″ across. Stems pulled down with weight of flower heads.

Leaves 3-6″ long, green ovals.

FALL COLOR: bronze to wine red.

Soil: moist and well drained; add peatmoss, leafmold or compost. Apply mulch in spring after soil has warmed, to retain moisture and feed the surface roots.

Light: best in partial shade; tolerates afternoon sun.

Prune back previous season's growth by about half, in late winter. For larger, fewer flower heads, cut back to 2 good buds. Remove dead flower heads in fall or early spring.

Uses: shrub border, specimen or accent, informal hedge planted at 2-3′ spacing.

BIG-LEAVED or FRENCH HYDRANGEA *Hydrangea macrophylla*

Zones 7 8 9 10

Rounded deciduous shrub, height 3-6′.

Flowers on last year's growth, July through September; rounded balls of sterile flowers (picture), or flatter clusters with edge of sterile flowers (lace-cap). White, or pink or blue depending on soil pH.
pH 6.0-6.5 results in pink flowers
pH 5.0-5.5 results in blue flowers

Leaves 6-8″ long broad ovals.

FALL COLOR: shades of purple and wine.

Soil: moist, well drained; add peatmoss, leafmold or compost. Apply spring mulch when soil warmed, to retain moisture and feed surface roots. Add iron chelate, sequestrene, or aluminum sulfate to alkaline soils.

Light: partial shade; avoid early morning sun.

TOLERATES coastal climates.

Prune after flowering; cut out succession of 2, 3 or older shoots to ground level each year, to promote strong new growth. Remove dead flower heads in fall or early spring.

Uses: specimen, in borders, foundation planting, forced in greenhouses for indoor plants.

PEEGEE (PG) HYDRANGEA *Hydrangea paniculata 'Grandiflora'*

Zones 5 6 7 8 9
(not southern Florida) and 10,
plus southern areas in zone 4.

Deciduous shrub with arching branches, easily trained to small tree. Very fast growing, will add 3-5′ in one season. Unpruned height reaches 15-20′.

Flowers on this year's growth, erect clusters (panicles) in August and September. Creamy-white blossoms turn pink when mature. The 18″ long panicles can be dried for winter decoration.

Leaves: 3-6″ ovals, turning from green to dull bronze in fall.

Soil: moist, well drained; add peatmoss, leafmold or compost. Apply spring mulch when soil warmed, to retain moisture and feed surface roots.

Prune in late winter, cutting back previous season's growth to about half. For fewer larger flower heads, prune back to 2 buds on each shoot. Remove dead heads in fall or early spring.

Uses: specimen or accent.

GOLDEN ST. JOHN'S WORT *Hypericum frondosum*

Zones 5 6 7 8 9
(not southern Florida) & 10

Decidous shrub, fast growing with dense and compact habit. Height and spread about 3′.

Flowers in summer, bright golden yellow blossoms 1-2″ across, July to mid-August.

Leaves 1-3″ long, blue green color; held late in fall.

Soil: almost any, well drained; tolerates poor, dry, sandy or alkaline conditions.

Light: full sun or light shade.

Pruning rarely needed. Cut off dead wood in early spring.

Uses: rock gardens, shrub border, foundation planting, bedding plant.

WINTERBERRY *Ilex verticillata*

Zones 5 6 7 8 9 10

Slow growing deciduous shrub, maturing to 10′ tall. Need both male and female plants for good fruit production.

Flowers in summer, inconspicuous.

Fruit: abundant red berries, rarely eaten by birds, persist to midwinter.

Leaves: 1½-2½″ long, green, turning yellow in fall.

Soil: almost any, prefers slightly acid conditions.

Tolerates moist position. Mulch rather than cultivate around plants.

Light: best in full sun; tolerates shade though bears fewer berries.

Pruning rarely needed. Trim hedge in summer.

Uses: specimen or accent, hedges.

GLOBEFLOWER or DOUBLE KERRIA *Kerria japonica 'Pleniflora'*

Zones 4 5 6 7 8 9 (not southern Florida) and 10.

Open irregular deciduous shrub, grows to 6′ tall if unpruned. Shoots characteristic green color all year. Fine textured shrub.

Flowers in spring, double yellow blossoms up to 2″ across.

Leaves wedge shaped, 1½-4″ long, mid green all summer.

FALL COLOR: yellow.

Soil: almost any; add peatmoss, leafmold or compost; tolerates poor soil conditions.

Light: best in partial shade.

Prune after flowering; cut back flowered shoots to strong newly growing shoots. Thin main shoots to encourage new ones. Trim off frost damaged shoots as new growth starts in early spring.

Uses: informal hedge or screen, specimen or accent, shrub border.

BEAUTY BUSH *Kolkwitzia amabilis*

Zones 4 5 6 7 8 9
(not southern Florida) & 10.

Deciduous shrub with long arching branches. Height and spread 6-10′. Bark characteristic tan to gray color; and peeling, lending plant shaggy appearance in winter.

Flowers in spring to early summer. Pink, yellow throated tubes, ½″ across, profusely clustered along branches.

Fruits persist to early winter; dry hairy brown seed cases.

Leaves 1-3″ long, gray-green in summer.

FALL COLOR: dull red.

Soil: almost any, well drained.

Light: full sun or light shade.

Prune only to thin out older stems, keeping mature plant vigorous.

Uses: specimen, shrub border.

CRAPE MYRTLE *Lagerstroemia indica*

Zones 7 8 9 10

Deciduous flowering shrub with varieties reaching heights of 15-20′.

Bark characteristic gray to tan and smooth, peeling off to reveal pinkish new bark.

Flowers in summer on current season's growth; blossoms up to 1½″ wide, with crinkled petals, in clusters 4-9″ long. Varieties in white, lavender, pinks and reds.

Leaves 1-2″ ovals; bronze in spring, turning green in summer.

FALL COLOR: yellow, orange, red.

Soil: moist and well drained; add peatmoss, leafmold or compost.

Light: best in full sun.

Prune as needed in early spring.

Uses: specimen or accent, hedge or screen, planted at 4′. Use for attractive stems.

GOLDEN VICARY PRIVET *Ligustrum ×vicaryi*

Zones 4 5 6 7 8 9
(not southern Florida) & 10

Densely branching deciduous shrub. Unpruned height reaches 10-12′.

Flowers in summer, small clusters of white blossoms, scented.

Fruit: ¼″ black berries, persist, seldom eaten by birds.

Leaves: 1-3″ long, smooth ovals, bright greenish gold throughout summer. Leaves hold to late fall.

Soil: almost any, well drained. Heavy feeder; add peatmoss, compost or leafmold.

Light: full sun or light shade.

TOLERATES urban pollution.

Pruning: clip or trim regularly as hedge. Remove dead or older branches in early spring, cutting right to base to maintain succession of new strong growth.

Uses: hedge or screen, planted at 12-18″. Each spring, cut back new growth to half until desired height is reached.

WINTER HONEYSUCKLE *Lonicera fragrantissima*

Zones 5 6 7 8 9 (not southern Florida) and 10.

Spreading semi-evergreen shrub. Height 6-8′; spread equals height.

Flowers in early spring, fragrant white to pink blossoms. Branches cut in midwinter will flower indoors.

Fruit: red berries mature in early summer.

Foliage: stiff and leathery, dark green ovals remaining throughout winter in milder climates.

Soil: almost any, moist and well drained.

Light: best in full sun.

TOLERATES urban pollution, poor soil, insect pests.

Pruning rarely needed, except occassional thinning. Clip regularly to maintain formal appearance.

Uses: hedge-formal or informal, specimen, screen (self-supporting).

STAR MAGNOLIA *Magnolia stellata*

Zones 7 8 9 10,
plus east areas in zones 4 5 and 6.

Multiple, single or trunks, densely growing deciduous shrub or tree. Slow growth rate. Height 6-10′ in 6-8 years, matures 15-20′. Starts flowering when about 3′ tall.

Flowers in very early spring before leaves unfurl. White 3″ fragrant star shaped blossoms.

Fruit knobbly and green, mature red in fall and open to reveal red inedible fruits.

Leaves 2½″ long and pointed, pale to mid-green turning bronze-yellow in fall.

Soil: deep, well drained, acid (pH 5.0-6.0); add peatmoss or compost. Plant no deeper than in previous position. Apply spring mulch though not touching bark.

Light: full sun, light shade in south. Shelter from winds.

TOLERATES urban pollution.

Pruning rarely needed.

Uses: specimen or accent.

GOLDEN MOCK ORANGE *Philadelphus coronarius 'Aureus'*

Zones 5 6 7 8 9
(not southern Florida) and 10.

Densely growing upright deciduous shrub. Height 6-8′. Easy to establish.

Flowers in spring and early summer, single 1-1½″ white blossoms. Fragrance like orange blossom.

Leaves: bright golden yellow in spring, turning green-yellow in summer.

Soil: prefers moist and well drained, with added peatmoss, compost or leafmold.

Light: full sun or light shade.

TOLERATES urban pollution, dry soil.

Prune immediately after flowering, cutting back flowered shoots to a strongly growing bud or new shoot. Thin out older wood, retain young shoots.

Uses: in shrub borders, group planting, specimen or accent.

MOCK ORANGE *Philadelphus ×virginalis*

Zones 5 6 7 8 9
(not southern Florida) and 10.

Fast growing deciduous shrub with slender erect branches. Height to 8′. Easy to establish.

Flowers in early summer, continues sparingly into fall. Orange blossom fragrance, pure white waxy flowers. Single, semi-double, or double, 1-2″ across; in clusters at ends of branches. Dry fruit.

Leaves oval and pointed, 1½-2½″ long, mid green.

Soil: prefers moist well drained soil with added peatmoss, leafmold or compost.

Light: full sun or light shade.

TOLERATES or resists attack by diseases and insect pests.

Prune immediately after flowering, cutting back flowered shoots to a strongly growing bud or new shoot. Thin out older wood, retain young shoots.

Uses: in shrub border, specimen, in group planting.

EASTERN NINEBARK *Physocarpus opulifolius 'Luteus'*

All zones

Fast growing deciduous shrub with irregular habit, reaches 9′ high if untrimmed.

Flowers in early summer, clusters of small white or pinkish-white blossoms.

Fruit: dry capsules ripening reddish to brown, persist through winter.

Leaves: bright yellow in spring, maturing yellow-green. Shrubs appear to be covered with yellow flowers, especially in spring when leaf color is brightest.

Soil: almost any, well drained.

Light: full sun or partial shade.

TOLERATES poor soil and hard winters.

Prune or trim to contain size.

Uses: informal hedge or screen; plant 2-3′ apart; fast growing filler for shrub border, behind specimen shrubs; best for large areas.

BUSH CINQUEFOIL *Potentilla fruticosa*

Zones 3 4 5 6 7 8 9 10

Spreading low shrub, deciduous in north and semi-evergreen in milder climates. Height 2-4′; spread 4-5′.

Flowers in early summer, continuing through fall. Blossoms about 1″ across, single and rose-like, light golden yellow and white varieties. Dry fruit.

Leaves pinnate with 3, 5 or 7 leaflets 1″ long. Color gray green, covered with silky hairs.

Soil: almost any, well drained; prefers light soil.

Light: full sun.

Pruning rarely needed. Remove old and dead wood, cutting back to ground level. After flowering trim off shoot tips to encourage strong branching.

Uses: borders, ground cover, plant in groups for greater effect.

PURPLE LEAF SAND CHERRY *Prunus ×cistena*

Zones 3 4 5 6 7 8 9 (not southern Florida) and 10.

Hardy deciduous shrub. Height about 8′, spread 4-6′.

Flowers in spring, white or light pink to contrast with foliage.

Leaves remain rich wine red all summer.

FALL COLOR: darker purple.

Soil: moist and well drained; some lime preferred. Shallow roots system. Plant no deeper than in previous position.

Light: best in full sun.

TOLERATES cold winters and hot summers of central and northern plains.

Pruning not needed. Shape by clipping in late summer or right after flowering.

Uses: specimen or accent, low hedge; can be sheared any time during growing season.

FLOWERING ALMOND *Prunus glandulosa*

Zones 4 5 6 7 8 9 (not southern Florida) and 10.

Low growing small tree or shrub. Height and spread 4-6′. Branches cut in late winter will flower indoors.

Flowers in early spring before leaves unfurl. Pink or white double blossoms about 1″ across. No fruit.

Soil: moist and well drained, some lime preferred. Shallow root system, plant no deeper than in previous position.

Light: best in full sun.

Prune immediately after flowering; cut off flowered portions of stems to encourage new growth for next year's flowers.

Uses: hedge or screen, back of border, accent or specimen, trained against wall or screen, patio.

MOLLIS HYBRID AZALEAS *Rhododendron ×kosterianum*

Zones 5 6 7 8 10

Deciduous flowering shrubs. Height to 5′, spread equals or exceeds height.

Flowers in late spring or early summer, before most of leaves unfurl. 2-3″ blossoms in yellows, oranges, pinks or white.

Leaves mid to dark green all summer, brownish in fall.

Soil: moist and well drained, acid (pH 4.5-5.5); add peatmoss, leafmold or compost. Surface root system.

Water well especially if dry during flowering. Mulch annually to retain moisture in summer.

Light shade; protect from heat of midday and afternoon sun. Deep shade or continued damp will cause mildew attack. Protect from winds.

Pruning rarely needed; remove dead or unsightly stems.

Uses: foundation planting, shrub borders, light woodland, accent.

MOLLIS HYBRID AZALEAS *Rhododendron ×kosterianum*

Zones 5 6 7 8

Deciduous flowering shrubs. Height to 5′, spread equals or exceeds height.

Flowers in late spring or early summer, before most of leaves unfurl. 2-3″ blossoms in yellows, oranges, pinks or white.

Leaves mid to dark green all summer, brownish in fall.

Soil: moist and well drained, acid (pH 4.5-5.5); add peatmoss, leafmold or compost. Surface root system.

Water well especially if dry during flowering. Mulch annually to retain moisture in summer.

Light shade; protect from heat of midday and afternoon sun. Deep shade or continued damp will cause mildew attack. Protect from winds.

Pruning rarely needed; remove dead or unsightly stems.

Uses: foundation planting, shrub borders, light woodland, accent.

MOLLIS HYBRID AZALEAS *Rhododendron* ×*kosterianum*

Zones 5 6 7 8

Deciduous flowering shrubs. Height to 5′, spread equals or exceeds height.

Flowers in late spring or early summer, before most of leaves unfurl. 2-3″ blossoms in yellows, oranges, pinks or white.

Leaves mid to dark green all summer, brownish in fall.

Soil: moist and well drained, acid (pH 4.5-5.5); add peatmoss, leafmold or compost. Surface root system.

Water well especially if dry during flowering. Mulch annually to retain moisture in summer.

Light shade; protect from heat of midday and afternoon sun. Deep shade or continued damp will cause mildew attack. Protect from winds.

Pruning rarely needed; remove dead or unsightly stems.

Uses: foundation planting, shrub borders, light woodland, accent.

ELDERBERRY *Sambucus canadensis*

Zones 4 5 6 7 8 9
(not southern Florida) & 10.

Fast growing deciduous shrub. Height to 12′ with equal spread. Grows 3′ or more each year when young. Excellent as large shrub in borders.

Flowers in late spring and early summer; small white blossoms in flat clusters 6-8″ across.

Fruit: small blue-black berries in clusters, edible (for wines, preserves, pies).

Leaves: coarse and compound with 5-9 leaflets.

Soil: moist and fertile, well drained.

Light: full sun.

Prune as desired in early spring.

Uses: moist, well drained soils, edges of light woodland, swampy places.

DWARF MOUNTAIN ASH *Sorbus tianshanica*

Zones 6 7 8, plus north and central areas in zone 10

Slow growing shrub or small tree. Height to 10-15′, spread 4-6′.

Flowers in late spring, clusters of white ¾″ blossoms.

Fruit ¼-½″ waxy bright red berries in large clusters, 3-5″ across.

Leaves dark green and lustrous, closely packed in groups of 11-14 leaflets.

FALL COLOR: orange red.

Soil: almost any, well drained. Short lived in alkaline conditions. Shallow root system. Mulch, but not touching trunk.

Light: best in full sun.

Prune young trees in winter or early spring.

Uses: accent, in shrub border, foundation planting, patio tree—useful in restricted areas.

ANTHONY WATERER SPIREA *Spiraea ×bumalda 'Anthony Waterer'* Zones 5 6 7 8 9 10

Low growing erect deciduous shrub. Height 2′ with spread equal to or greater than height.

Flowers in midsummer on this year's growth. Rose pink blossoms in flat clusters.

Leaves long and narrow with serrated margins. Pinkish green when young, turning mid green; some variegated leaves.

Soil: almost any.

Light: best in full sun.

TOLERATES moist conditions.

Prune in early spring to encourage new growth for flowers in the same year.

Uses: in shrub borders, low hedge, specimen.

BRIDAL WREATH *Spiraea prunifolia*

Zones 4 5 6 7 8 9 10

Deciduous shrub with erect gracefully curving stems. Height 5-8′.

Flowers on last year's growth, before leaves unfurl in spring. White blossoms line slender stems.

Leaves dark green and glossy.

FALL COLOR: brilliant oranges and reds.

Soil: almost any.

Light: best in full sun.

Prune immediately after flowering. Cut back flowered stems to strong new growth; remove older branches entirely to encourage succession of new basal shoots.

Uses: specimen or accent, in shrub borders, background.

VANHOUTTE SPIREA *Spiraea ×vanhouttei*

Zones 4 5 6 7 8 9 10

Deciduous shrub dense growing with long arching stems. Grows to height of 6-8′.

Flowers on last year's growth, in late spring. Clusters of tiny white blossoms hide young leaves.

Leaves: dark green pointed ovals.

FALL COLOR: orange red.

Soil: almost any.

Light: best in full sun.

Prune immediately after flowering. Cut back or shear flowered stems; remove older branches entirely to encourage succession of new basal shoots.

Uses: hedge, plant 15-24″ apart and cut back once to 6″ from the ground. Tip subsequent shoots to encourage branching; shear after flowering each. Shrub borders, background.

PERSIAN LILAC *Syringa ×persica*

Zones 5 6 7 8 9 10

Dense upright deciduous shrub with gracefully arching stems. Height and spread 5-6′.

Flowers in mid to late spring, very fragrant blossoms in small clusters grouped at stem tips. Followed by unattractive dry fruits.

Leaves oval and pointed, dark green all season.

Soil: almost any, well drained; best if near neutral (pH 7.0); tolerates poor soil conditions. Add peatmoss or leafmold.

Light: best in full sun.

Prune right after flowering; cut out dead flower heads and remove older and weaker stems by pruning to ground level.

Uses: specimen or accent, screen or hedge, plant 2-4′ apart.

COMMON or FRENCH LILAC *Syringa vulgaris*

Zones 4 5 6 7 8

Fast growing deciduous shrub with arching branches. Height 8-15′; spread 5-10′.

Flowers in spring, very fragrant single or double flowers in clusters 6-8″ long. Varieties in color from white to purples, pinks, and yellow. Unattractive dry fruit clusters.

Leaves: smooth edged pointed ovals, mid green color all season.

Soil: almost any, well drained; best if near neutral (pH 7.0); tolerates poor soil conditions. Add peatmoss or leafmold.

Light: best in full sun.

TOLERATES urban pollution.

Prune right after flowering; cut out dead flower heads and remove older and weaker stems by pruning to ground level.

Uses: specimen, screen or hedge, plant 2-4′ apart.

COMMON or FRENCH LILAC *Syringa vulgaris*

Zones 4 5 6 7 8

Fast growing deciduous shrub with arching branches. Height 8-15′; spread 5-10′.

Flowers in spring, very fragrant single or double flowers in clusters 6-8″ long. Varieties in color from white to purples, pinks, and yellow. Unattractive dry fruit clusters.

Leaves: smooth edged pointed ovals, mid green color all season.

Soil: almost any, well drained; best if near neutral (pH 7.0); tolerates poor soil conditions. Add peatmoss or leafmold.

Light: best in full sun.

TOLERATES urban pollution.

Prune right after flowering; cut out dead flower heads and remove older and weaker stems by pruning to ground level.

Uses: specimen, screen or hedge, plant 2-4′ apart.

COMMON or FRENCH LILAC *Syringa vulgaris*

Zones 4 5 6 7 8

Fast growing deciduous shrub with arching branches. Height 8-15′; spread 5-10′.

Flowers in spring, very fragrant single or double flowers in clusters 6-8″ long. Varieties in color from white to purples, pinks, and yellow. Unattractive dry fruit clusters.

Leaves: smooth edged pointed ovals, mid green color all season.

Soil: almost any, well drained; best if near neutral (pH 7.0); tolerates poor soil conditions. Add peatmoss or leafmold.

Light: best in full sun.

TOLERATES urban pollution.

Prune right after flowering; cut out dead flower heads and remove older and weaker stems by pruning to ground level.

Uses: specimen, screen or hedge, plant 2-4′ apart.

COMMON or FRENCH LILAC *Syringa vulgaris*

Zones 4 5 6 7 8

Fast growing deciduous shrub with arching branches. Height 8-15′; spread 5-10′.

Flowers in spring, very fragrant single or double flowers in clusters 6-8″ long. Varieties in color from white to purples, pinks, and yellow. Unattractive dry fruit clusters.

Leaves: smooth edged pointed ovals, mid green color all season.

Soil: almost any, well drained; best if near neutral (pH 7.0); tolerates poor soil conditions. Add peatmoss or leafmold.

Light: best in full sun.

TOLERATES urban pollution.

Prune right after flowering; cut out dead flower heads and remove older and weaker stems by pruning to ground level.

Uses: specimen, screen or hedge, plant 2-4′ apart.

COMMON or FRENCH LILAC *Syringa vulgaris*

Zones 4 5 6 7 8

Fast growing deciduous shrub with arching branches. Height 8-15′; spread 5-10′.

Flowers in spring, very fragrant single or double flowers in clusters 6-8″ long. Varieties in color from white to purples, pinks, and yellow. Unattractive dry fruit clusters.

Leaves: smooth edged pointed ovals, mid green color all season.

Soil: almost any, well drained; best if near neutral (pH 7.0); tolerates poor soil conditions. Add peatmoss or leafmold.

Light: best in full sun.

TOLERATES urban pollution.

Prune right after flowering; cut out dead flower heads and remove older and weaker stems by pruning to ground level.

Uses: specimen, screen or hedge, plant 2-4′ apart.

TAMARISK *Tamarix ramosissima (T. pentandra)*

All zones

Multistemmed tree or shrub growing 12-15′ tall. Long branches, graceful and fine texture.

Flowers in mid to late summer, fluffy pink blossoms in spikes at the ends of branches.

Leaves ¼″ long, scale-like, clinging to stems. Pale green with blue-gray overcast.

Soil: lime-free, well drained. Root system coarse and brittle.

Light: full sun.

TOLERATES coastal climate, desert conditions, wind.

Prune in late winter or early spring; remove all but 2-3″ of previous season's growth, to encourage branching and new shoots for flowers this year.

Uses: specimen or accent, windbreak, plant 24″ apart and cut back once to 12″ high. Tip new shoots when they are 6″ long.

FRAGRANT SNOWBALL *Viburnum carlcephalum* Zones 5 6 7 8 9 (not southern Florida) and 10

Deciduous shrub, makes a dense bush 6-8′ high.

Flowers in spring; pinkish buds open to rounded 4-5″ clusters of fragrant white fertile blossoms.

Fruit: red berries ripen black in midsummer; fruit tends to be hidden by new growth.

Leaves 3-4″ long, dark green.

FALL COLOR: purplish red to bronze.

Soil: almost any, well drained. Tolerates moist conditions.

Light: full sun or light shade.

Prune to remove dead or weak, overcrowded growth as necessary.

Uses: specimen or accent, informal screen or hedge, planted 4-6' apart.

ARROWWOOD *Viburnum dentatum*

Zones 3 4 5 6 7 8 9 (not southern Florida) and 10.

Vigorous deciduous shrub with long arching branches. Height 10-15′.

Flowers in early summer, 2″ clusters of tiny white fertile blossoms.

Fruit: dark blue berries in early fall. Persist if not eaten by birds.

Leaves: glossy green, coarsely toothed, 2-4″ long.

FALL COLOR: red to bronze.

Soil: almost any, well drained; tolerates moist conditions.

Light: full sun or light shade.

Prune only to remove dead or weak growth, cutting stems right back to their origin to retain natural arching shape.

Uses: specimen or accent, in shrub borders, group planting. Hedge or screen.

WAYFARING TREE *Viburnum lantana*

Zones 3 4 5 6 7 8 9 (not southern Florida) and 10.

Fast growing deciduous shrub or small tree. Matures 10-15′ tall. Easy to grow.

Flowers in spring. 2-4″ flat clusters of tiny white blossoms.

Fruit: red berries turning black in early fall. Persist if not eaten by birds.

Leaves rounded, gray-green, 2-5″ across.

FALL COLOR: red.

Soil: almost any, well drained. Prefers moist soil.

Light: full sun or light shade.

Prune only to remove dead wood as needed. Trim for formal appearance.

Uses: specimen or accent, in woodland groups, informal plantings.

DWARF EUROPEAN CRANBERRY BUSH *Viburnum opulus 'Nanum'* Zones 3 4 5 6 7 8 9 (not southern Florida) and 10.

Small densely leaved deciduous shrub. Height 2-2½′. Easy to grow.

Sterile—Seldom flowers or fruits.

Leaves tri-lobed, dark green.

FALL COLOR: red.

Soil: almost any, well drained. Prefers moist conditions; tolerates heavy clay soils.

Light: full sun or light shade.

Prune only to remove dead wood as needed. Trim to maintain shape in formal hedge.

Uses: low hedge, plant 2-3′ apart.

SNOWBALL BUSH *Viburnum opulus 'Roseum'*

Zones 3 4 5 6 7 8 9 (not southern Florida) and 10

Fast growing deciduous shrub, height 9-12′. Easy to grow.

Flowers in late spring, round clusters of white sterile florets.

Leaves trilobed, dark green.

FALL COLOR: red.

Soil: almost any, well drained. Prefers moist conditions, tolerates heavy clay soils.

Light: full sun or light shade.

TOLERATES urban pollution.

Prune only to remove dead wood as needed. Trim to shape in hedge or screen.

Uses: hedging or screening, spaced 3-6′ apart, light woodland, shrub border.

DOUBLEFILE VIBURNUM *Viburnum plicatum var. tomentosum*

Zones 4 5 6 7 8 9 (not southern Florida) and 10.

Deciduous shrub with long horizontal and arching branches. Matures 5-7′ tall with equal spread.

Flowers in late spring, flat clusters on tiny white fertile blossoms encircled with larger white sterile flowers.

Fruit: clusters of red berries turn black in midsummer. Persist if not eaten by birds.

Leaves 2-4″ long, oval and pointed. Dark green with whitish undersides.

FALL COLOR: purplish red to bronze.

Soil: almost any, well drained; tolerates moist conditions.

Light: full sun or light shade.

Prune only to remove dead or weak growth, cutting stems right back to their origin to retain long arching natural appearance.

Uses: specimen or accent, hedge, plant 3-6′ apart.

S 379

SIEBOLD VIBURNUM *Viburnum sieboldii*

Zones 4 5 6 7 8 9 (not southern Florida) and 10.

Fast growing deciduous shrub or small tree. Matures 6-10′ tall. Easy to grow.

Flowers in late spring, 3-4″ clusters of tiny cream-white blossoms.

Fruit: clusters of green berries, maturing red, dark blue and black by early fall. Persist if not eaten by birds.

Leaves 2-5″ long, coarsely toothed, dark green and glossy.

FALL COLOR: rich red.

Soil: almost any, well drained. Tolerates moist conditions.

Light: full sun or light shade.

Prune if needed to remove dead wood. Trim to maintain formal shape.

Uses: specimen or accent.

CHASTE TREE *Vitex agnus-castus*

Zones 6 7 8 9 10

Rapidly growing deciduous shrub. If killed to ground in winter, will reach 3-5′ by midsummer. In milder climates height reaches 20′; spread equals height. Grows faster in hot weather.

Flowers in midsummer and early fall; aromatic tiny blue or white blossoms in dense spikes.

Leaves 3-4″ long in groups of 5-7; gray green and aromatic.

Soil: almost any, well drained.

Light: full sun.

Prune in spring to remove winter killed branches. Often killed to ground in zone 6.

Uses: specimen or accent.

WEIGELA *Weigela florida and hybrids*

Zones 5 6 7 8 9
(not southern Florida) and 10.

Rapidly growing deciduous shrub with arching branches. Height 6-10′; spread equals or exceeds height.

Flowers in June on previous year's growth; clusters of wide-lipped tubular blossoms. Varieties pink or red.

Leaves about 4″ long, mid to dark green. May be purplish green or edged with creamy-white.

Soil: almost any; prefers dry conditions.

Light: best in full sun.

Resists attack by insect pests.

Prune in spring to remove winter killed twigs. After flowering, trim back to good bud or new shoot. Cut back older stems right to ground to maintain succession of new growth.

Uses: Excellent in shrub borders.

Flowering Trees

RUBY HORSECHESTNUT *Aesculus ×carnea 'Briottii'*

Zones 3 4 5 6 7 8 9 10

Round headed deciduous tree. Height 20′ in 10-15 years, matures to 30-40′; spreads 20′. Starts flowering on 6-8′ young tree.

Flowers in May, brilliant red 5-7″ long spikes.

Fruit are spiked capsules containing non-edible brown nuts (poisonous).

Leaves open in early spring, palmate, glossy dark green. Usually no fall color.

Soil: moist and fertile; deep root system.

Light: full sun or light shade.

TOLERATES urban pollution and coastal climates.

Prune if required, in early spring.

Uses: shade, street, accent tree.

SILK TREE or MIMOSA *Albizia julibrissin*

Zones 6 7 8 9 10

Flat topped spreading deciduous tree. Very fast growing with single or multiple trunks. Height 15-20′ in 3-5 years, matures to 25-35′; spread equals height. Starts flowering at 2-3 years, loses vigor in 25-30 years.

Flowers in summer, pink clusters above foliage. Followed by 4-6″ long flat seed pods.

Leaves pinnate, feathery and acacia-like. Casts gentle shade that does not inhibit grass.

Soil: almost any, well drained; tolerates poor soil including alkaline conditions.

Light: best in full sun.

TOLERATES wind. In colder areas of recommended zones, wrap stems with burlap or heavy paper for winter protection in early years.

Prune as desired in winter or early spring.

Uses: accent, lawn specimen, small shade tree, good for small areas, patio.

SERVICEBERRY or JUNEBERRY *Amelanchier canadensis*

Zones 4 5 6 7 8,
plus north in zone 10

Open oval headed deciduous tree. Fast growing with single or multiple trunks. Height 15-18′ in 5-7 years, matures to 20-30′; spreads 10-15′. Starts flowering in first year. Smooth light gray bark.

Flowers before leaves unfurl in very early spring. White blossoms followed by tiny red-purple edible fruit, in early summer.

Leaves mid green, 1½-3″ long.

FALL COLOR: yellow, orange, red.

Soil: moist yet well draining; supply water in excessively dry periods.

Light: full sun or light shade.

Rarely needs pruning.

Uses: accent, group plantings, fine at edge of informal woodland group.

EASTERN REDBUD or JUDAS TREE *Cercis canadensis*

Zones 4 5 6 7 8 9 10

Irregular spreading deciduous tree or shrub. Moderate growth rate. Height 10-15′ in 10-12 years, mature to 35′; spread equal to or greater than height. Starts flowering at 4-5 years. Taproot makes it hard to transplant trees over 6′ tall.

Flowers before leaves unfurl in early spring; purplish pink blossoms, followed by 2-4″ long greenish pink seedpods which dry brown and persist into winter.

Leaves open in late spring, heart-shaped, 3-5″ long, green.

FALL COLOR: yellow.

Soil: rich sandy loam, deep and very well drained.

Light: full sun or light shade.

Pruning: rarely needed.

Uses: accent, in group or informal plantings.

FRINGE TREE *Chionanthus virginicus*

Zones 4 5 6 7 8 9 10

Slow growing deciduous tree or large shrub with single or multiple trunks. A 4-5′ tree grows to 12-15′ with equal spread in 10-15 years, matures to 25-30′. Starts flowering when 2-4′ tall.

Flowers in late spring or early summer; 6-8″ pendulous clusters of minute individual fragrant white blossoms.

Fruit (female trees only): hanging clusters of dark blue berries with bitter taste.

Leaves open very late in spring, 3-8″ long.

FALL COLOR: bright golden yellow.

Soil: moist, deep and well drained; deep root system.

Light: full sun or light shade.

Pruning: rarely needed.

Uses: specimen or accent.

FLOWERING DOGWOOD *Cornus florida*

Zones 5 6 7 8

Spreading flat topped deciduous tree with single or multiple trunks. Moderate growth rate. Matures to 15-30′ high, spreads to 15-20′.

Bark characteristic nearly black and deeply furrowed on mature trees, alligator-like.

Flowers in early spring before leaves unfurl. Individual white bracts 3-5″ across. Followed by shiny red fruits which are eaten by birds.

Leaves: pinkish green when young, turning mid green in summer.

FALL COLOR: red.

Soil: moist, well drained, acid (pH 5.5-6.5). Supplement with leafmold, peatmoss or compost. Mulch in spring after soil has warmed. Water thoroughly in dry periods.

Light: full sun in north, light shade in south.

Pruning not usually recommended since dogwoods are slow to heal.

Uses: specimen or group.

FLOWERING DOGWOOD *Cornus florida 'Rubra'*

Zones 5 6 7 8

Spreading flat topped deciduous tree with single or multiple trunks. Moderate growth rate. Matures to 15-30′ high, spreads to 15-20′.

Bark characteristic nearly black and deeply furrowed on mature trees, alligator-like.

Flowers in early spring before leaves unfurl. Individual bracts 3-5″ across, white or pink. Followed by shiny red fruits which are eaten by birds.

Leaves: pinkish green when young, turning mid green in summer.

FALL COLOR: red.

Soil: moist, well drained, acid (pH 5.5-6.5). Supplement with leafmold, peatmoss or compost. Mulch in spring after soil has warmed. Water thoroughly in dry periods.

Light: full sun in north, light shade in south.

Pruning not usually recommended since dogwoods are slow to heal.

Uses: specimen or group.

JAPANESE (KOUSA) DOGWOOD *Cornus kousa*

Zones 5 6 7 8

Spreading small deciduous tree. Moderate growth rate. Matures to 15-25′ high, spreads to 12-20′.

Flowers in late spring and early summer after leaves have expanded. Flat sprays of 3-5″ blossoms along tops of branches; white or tinged pink. Flowers followed by soft red berries, eaten by birds.

Leaves: strong green color, up to 3″ long.

FALL COLOR: reddish brown.

Soil: moist, well drained, acid (pH 5.5-6.5). Supplement with leafmold, peatmoss or compost. Mulch every spring after soil has warmed. Water thoroughly when dry, particularly in late summer when new buds forming.

Light: full sun in north, light shade in south.

Pruning not usually recommended since dogwoods are slow to heal.

Uses: specimen or accent, in group plantings.

HAWTHORN or COCKSPUR THORN *Crataegus crus-galli*

Zones 4 5 6 7 8,
plus north in zone 10.

Spreading deciduous tree with dense horizontal branches persisting to ground level if unpruned. Curving thorns 1½-3″ long*. Moderate growth rate. Height 20′ in 10-15 years, matures to 25-35′. Easily trained to shape for use as hedge or screen. Starts flowering about 2 years after planting. Taproot makes it hard to transplant trees over 8′ tall.

Flowers in May and June, white clusters along branches. Small red fruits persist through winter.

Leaves: 1½-3″ long, glossy green.

FALL COLOR: orange and red.

Soil: almost any, well drained; tolerates acid or alkaline conditions, prefers heavy or organic soils.

Light: best in full sun.

TOLERATES urban pollution, cold, drought, waterlogging.

Prune during winter to shape top. Remove basal suckers at any time.

Uses: accent, street tree, roadside planting, hedge or barrier, shade.

*The Unarmed, or thornless, Cockspur Thorn has none.

ENGLISH HAWTHORN *Crataegus laevigata 'Paulii (C. oxycantha)*

Zones 4 5 6 7 8, plus north in zone 10

Thickly branched deciduous tree with ½-1″ thorns. Moderate growth rate. Height to 25′; easily trained to shape in hedge or barrier. Starts flowering about 2 years after planting. Taproot makes it hard to transplant trees over 8′ tall.

Flowers in spring; clusters of pink, red or white blossoms crowd along branches; varieties single or double flowered. Small red fruits follow single flowers, persist through winter.

Leaves: 3 or 5 lobed, up to 2½″ across; mid to dark green turning yellow in fall.

Soil: almost any, well drained; tolerates acid or alkaline conditions, prefers heavy or organic soils.

Light: best in full sun.

TOLERATES urban pollution, cold, drought, waterlogging.

Prune during winter to shape top. Remove basal suckers any time.

Uses: accent, hedge or barrier, shade, street tree.

WASHINGTON HAWTHORN *Crataegus phaenopyrum (C. cordata)*

Zones 4 5 6 7 8,
plus north in zone 10

Spreading deciduous tree with thick horizontal branches and 1½-3″ thorns. Moderate growth rate. Height 15-20′ in 10-15 years, matures to 25-30′. Starts flowering about 2 years after planting. Tap root makes it hard to transplant trees over 8′ tall.

Flowers in spring, white clusters massed along branches. Numerous small red fruits persist through winter.

Leaves glossy green, ½-3″ long.

FALL COLOR: oranges and reds.

Soil: almost any, well drained; tolerates acid or alkaline conditions, prefers heavy or organic soils.

Light: best in full sun.

TOLERATES urban pollution, cold, drought, waterlogging.

Prune during winter to shape top. Remove basal suckers any time.

Uses: accent or specimen, hedge or barrier, shade tree, street tree.

FRANKLIN TREE *Franklinia alatamaha*

Zones 7 8 9 (not southern Florida) and 10.

Shrub or tree growing to 30′ high in mild climates; matures to 10′ only in northern areas in recommended zones. Striped smooth bark on older specimens.

Flowers in late summer and fall; pure white frilled petals and golden stamens in 3″ cup shaped blossoms, cover plant until frost.

Leaves: 5-6″ long, glossy dark green with silvery reverse.

FALL COLOR: bright scarlets and oranges. Best color in full sun.

Soil: moist and well drained, prefers acid soil; add peatmoss or compost.

Light: full sun or light shade.

Uses: specimen or accent.

No pruning needed.

GOLDEN RAIN TREE *Koelreuteria paniculata*

Zones 5 6 7 8 9 10

Deciduous tree with moderate growth rate; an 8-10′ tree grows to 20′ in 5-10 years, matures to 20-30′; spreads 10-20′.

Flowers in early summer, 12-15″ open large clusters of small yellow blossoms.

Fruit: 2″ bladder-like seed pods, first green, turning pink then brown when ripe.

Leaves 8-15″ compound. Reddish in spring, maturing blue green in summer, yellow in fall.

Soil: almost any, well drained; tolerates alkaline conditions.

Light: full sun.

Prune young trees in spring; mature trees rarely need pruning.

Uses: shade tree, specimen or accent, street tree, patio.

SOUTHERN MAGNOLIA or BULL BAY *Magnolia grandiflora*

Zones 8 9 10

Evergreen tree growing to 10-15′ in about 10 years; spreads 6-10′. Matures to 90′ in 75-100 years. Seedlings first flower at 15 years, grafted trees at 2-3 years old.

Flowers in spring and summer, a few into fall where night temperatures are above 40°F. Large 8-12″ blossoms, creamy white and fragrant.

Leaves: 5-8″ long, waxy and glossy dark green with tan-colored feltlike undersides.

Soil: moist and well drained.

Light: full sun if soil is moist; tolerates as little as 3 hours of sun per day in dryer conditions.

TOLERATES insect pests.

Pruning rarely needed.

Uses: specimen or accent, shade tree, street tree, espaliered on wall or screen.

SAUCER MAGNOLIA *Magnolia ×soulangiana*

Zones 7 8 9 10, plus east areas in zones 4 5 and 6.

Single or multiple trunks, deciduous tree or shrub. Moderate growth rate. Height 15-20′ in 10-15 years, matures 20-25′; spread equal to or exceeds height. Smooth gray bark. Starts flowering when about 4′ high.

Flowers in early spring before leaves unfurl. Candle-shaped buds open to large rose tinted white blossoms, fragrant.

Fruit knobbly and green, mature red in fall and open to reveal red inedible seeds.

Leaves 6″ long, broad ovals. Mid-green with slightly downy undersides, turn brown in fall.

Soil: deep, well drained, acid (pH 5.0-6.0): add peatmoss or compost. Plant no deeper than in previous position. Apply spring mulch though not touching bark.

Light: full sun, light shade in south. Shelter from winds.

Pruning rarely needed.

Uses: specimen or accent.

ALMEY CRAB APPLE *Malus hybrid 'Almey'*

Zones 4 5 6 7 8,
plus north in zone 10

Fairly rapidly growing flowering tree, spread exceeds height. Matures to 12-15′ high. Starts flowering in first or second season after planting.

Flowers in early spring as leaves unfurl; carmine-colored buds open to large crimson or pink flowers with white star at base of petals.

Fruit ripen maroon over orange base color.

Leaves: pinkish purple when young, turning glossy green in summer.

Soil: moist and well drained, acid (pH 5.0-6.5); add peatmoss, leafmold or compost. Shallow root system.

Stake newly planted trees until established (2-4 years).

Light: full sun.

Prune young trees in winter or early spring.

Uses: specimen or in group planting, roadside tree.

DOLGO CRAB APPLE Malus hybrid 'Dolgo'

Zones 3 4 5 6 7 8,
plus north in zone 10

Fast growing flowering tree; a 6-8′ tree will reach 15′ in 5-8 years, maturing to 35-40′. Naturally well shaped. Starts flowering in first or second season after planting. Tends to flower and fruit heavily in alternate years.

Flowers in very early spring as leaves unfurl; white, appleblossom scented blossoms, 1¾″ across.

Fruit: 1¼″ brilliant red, edible (very good for jellies).

Leaves: pinkish green when young, turning glossy dark green in summer.

Soil: moist and well drained, acid (pH 5.0-6.5); add peatmoss, leafmold or compost. Shallow root system.

Stake newly planted trees until established (2-4 years).

Light: full sun.

RESISTS cold.

Prune young trees in winter or early spring.

Uses: specimen or in group planting, fruit production.

DOROTHEA CRAB APPLE Malus hybrid 'Dorothea'

Zones 4 5 6 7 8,
plus north in zone 10

Fast growing flowering tree; a 6-8′ tree will reach 15′ in 8-10 years, maturing to 20-25′. Upright when young, spreading as tree matures. Starts flowering in first or second season after planting.

Flowers in early spring as leaves unfurl; semi double or double non-fading pink, scented blossoms, 1½-2″ across.

Fruit: ½″ bright yellow, edible.

Leaves: purplish green, when young, turning glossy green in summer.

Soil: moist and well drained, acid (pH 5.0-6.5); add peatmoss, leafmold or compost. Shallow root system.

Stake newly planted trees until established (2-4 years).

Light: full sun.

Prune young tree in winter or early spring.

Uses: specimen or in group planting, roadside tree.

HOPA CRAB APPLE *Malus hybrid 'Hopa'*

Zones 3 4 5 6 7 8,
plus north in zone 10

Fast growing flowering tree; a 6-8′ tree will reach 15′ in 5-8 years. Upright when young, spreading as tree matures. Starts flowering in first or second season after planting. Tends to flower and fruit heavily in alternate years.

Flowers in early spring as leaves unfurl; crimson buds open to single bright rose pink, scented blossoms.

Fruit: small red, edible (very good for jellies).

Leaves: bronze green throughout summer.

Soil: moist and well drained, acid (pH 5.0-6.5); add peatmoss, leafmold or compost. Shallow root system.

Stake newly planted trees until established (2-4 years).

Light: full sun.

RESISTS cold.

Prune young trees in winter or early spring.

Uses: specimen or in group planting, fruit production.

RADIANT CRAB APPLE *Malus hybrid 'Radiant'*

Zones 4 5 6 7 8,
plus north in zone 10

Fast growing flowering tree; a 6-8′ tree will reach 15′ in 5-8 years, maturing to 25-30′. Compact and symmetrical growth. Starts flowering in first or second season after planting.

Flowers in early spring as leaves unfurl; double bright rosy red, scented blossoms, 1″ across.

Fruit: ½″ crimson, edible.

Leaves: purple red when young, turning rich glossy green in summer.

Soil: moist and well drained, acid (pH 5.0-6.5); add peatmoss, leafmold or compost. Shallow root system.

Stake newly planted trees until established (2-4 years).

Light: full sun.

Prune young trees in winter or early spring.

Uses: specimen or in group planting, roadside tree.

RED SILVER CRAB APPLE *Malus hybrid 'Red Silver'*

Zones 4 5 6 7 8,
plus north in zone 10

Fast growing upright flowering tree; a 6-8′ tree will reach 15′ in 5-10 years, maturing to 25-30′. Starts flowering in first or second season after planting.

Flowers in early spring as leaves unfurl; red, scented blossoms, 1¼″ across.

Fruit: maroon, edible.

Leaves: purplish red with silver downy hairs when young, turning green in summer.

Soil: moist and well drained, acid (pH 5.0-6.5); add peatmoss, leafmold or compost. Shallow root system.

Stake newly planted trees until established (2-4 years).

Light: full sun.

Prune young trees in winter or early spring.

Uses: specimen or in group planting, background to green-leaved crab apples, roadside tree.

ROYALTY CRAB APPLE *Malus hybrid 'Royalty'*

Zones 4 5 6 7 8,
plus north in zone 10

Fast growing flowering tree; a 6-8′ tree will reach 10-12′ in about 6 years, maturing to approx. 15′. Starts flowering in first or second season after planting.

Flowers in early spring as leaves unfurl; crimson, scented blossoms, 1″ across.

Fruit: ½″ red, edible.

Leaves: rich glossy purple all summer.

Soil: moist and well drained, acid (pH 5.0-6.5); add peatmoss, leafmold or compost. Shallow root system.

Stake newly planted trees until established (2-4 years).

Light: full sun.

Prune young trees in winter or early spring.

Uses: specimen or in group planting, roadside tree.

VANGUARD CRAB APPLE *Malus hybrid 'Vanguard'*

Zones 3 4 5 6 7 8,
plus north in zone 10

Fast growing semi dwarf flowering tree; a 6-8′ tree will reach 10′ in 5-8 years. Pyramidal, upright habit. Starts flowering in first or second season after planting.

Flowers in early spring as leaves unfurl; large, bright rose-red, scented blossoms.

Fruit: small, bright red, edible.

Leaves: red when young, turning red-green in summer.

Soil: moist and well drained, acid (pH 5.0-6.5); add peatmoss, leafmold or compost. Shallow root system.

Stake newly planted trees until established (2-4 years).

Light: full sun.

RESISTS cold.

Prune young trees in winter or early spring.

Uses: specimen or in group planting, roadside tree, patio, small garden.

ELEYI CRAB APPLE *Malus purpurea 'Eleyi'*

Zones 4 5 6 7 8,
plus north in zone 10.

Fast growing flowering tree, spreading habit; a 6-8′ tree will reach 12′ in 5-8 years. Starts flowering in first or second season after planting.

Flowers in early spring as leaves unfurl; large single pink, scented blossoms.

Fruit: 1″ or larger, conical, dark red, edible.

Leaves: red purple to dark green all season.

Soil: moist and well drained, acid (pH 5.0-6.5); add peatmoss, leafmold or compost. Shallow root system.

Stake newly planted trees until established (2-4 years).

Light: full sun.

Prune young trees in winter or early spring.

Uses: specimen or in group planting.

SARGENT'S CRAB APPLE *Malus sargentii and Malus sargentii 'Rosea' (Pinkbud Sargent)* Zones 4 5 6 7 8, plus north in zone 10

Dwarf flowering tree, rarely taller than 6-8′ unless grafted onto standard stock. Low branching. Zig-zag branches may spread to 20′. Starts flowering in first or second season after planting.

Flowers profusely in early spring as leaves unfurl; faintly pink (rose pink in *Malus sargentii rosea*) buds open to pure white, scented blossoms, ½″ across.

Fruit: ¼″ dark red.

Leaves: pinkish green when young, turning strong green in summer.

Soil: moist and well drained, acid (pH 5.0-6.5); add peatmoss, leafmold or compost. Shallow root system.

Stake newly planted trees until established (2-4 years).

Light: full sun.

Prune young trees in winter or early spring.

Uses: specimen or in group planting, patio or small area.

SCHEIDECKER CRAB APPLE *Malus scheideckeri*

Zones 4 5 6 7 8,
plus north in zone 10

Fast growing flowering tree; a 6-8′ tree will reach 15′ in 5-8 years, maturing to 18-20′. Upright and compact habit. Starts flowering in first or second season after planting. Tends to flower and fruit heavily in alternate years.

Flowers in early spring as leaves unfurl; semi double, rose-pink, scented blossoms, 1½″ across.

Fruit: yellow-orange, edible.

Leaves: green-pink when young, turning glossy green in summer.

Soil: moist and well drained, acid (pH 5.0-6.5); add peatmoss, leafmold or compost. Shallow root system.

Stake newly planted tree until established (2-4 years).

Light: full sun.

Prune young trees in winter or early spring.

Uses: specimen or in group planting, roadside tree.

PURPLE LEAF FLOWERING PLUM *Prunus cerasifera 'Pissardi'*

Zones 4 5 6 7 8 9 10

Small deciduous tree with bushy round head. Grows to 15-20′ height, spread 10-15′.

Flowers pink in bud, opening to single white or light pink blossoms in spring. Few red fruit, eaten by birds.

Leaves unfurl with or right after flowers. Brilliant bronze-red color all summer.

FALL COLOR: brighter purple.

Soil: moist and well drained; some lime preferred. Shallow root system.

Light: full sun; protect from wind. Stake newly planted trees until established (2-4 years).

Pruning rarely needed.

Uses: specimen or accent, patio tree.

FLOWERING PEACH *Prunus persica*

Zones 6 7 8 9

Small deciduous tree, fast growing but short lived (10-15 years). Requires pruning for flower production. Height to 10′ with spread depending on pruning.

Flowers prolifically in early spring before leaves unfurl. Red, pink or white, usually double blossoms.

Leaves bright green all summer.

Soil: moist and well drained; some lime preferred. Shallow root system.

Light: full sun; protect from wind. Stake newly planted trees until established (2-4 years).

Prune back half the length of flowered shoots immediately after flowering.

Uses: specimen or accent, patio tree.

JAPANESE FLOWERING CHERRY *Prunus serrulata 'Kwanzan'*

Zones 6 7 8
plus north in 10

Medium sized deciduous tree, matures to 25-35′. Erect branches spread as tree matures. Starts flowering first season after planting.

Flowers prolifically in early spring, same time as new leaves unfurling. Deep pink double blossoms. No fruit.

Leaves: reddish when young, turning dark shiny green in summer.

Soil: moist and well drained; some lime preferred. Shallow root system.

Light: full sun, protect from wind. Stake newly planted trees until established (2-4 years).

Pruning rarely required.

Uses: accent or specimen, patio tree, in group planting.

WEEPING HIGAN CHERRY *Prunus subhirtella 'Pendula'*

Zones 6 7 8 and north in 10

Pendulous or weeping deciduous tree, grafted to standard rootstock to provide base height; this is usually about 6′; tree then matures to 20-30′; spread equals height.

Flowers in first season, cascades of light pink single or double blossoms line branches before leaves unfurl. Single flower varieties may have small cherries following flowers.

Leaves open after flowers in spring, mid green all season.

Soil: moist and well drained; some lime preferred. Shallow root system.

Light: full sun; protect from wind. Stake newly planted trees until established (2-4 years).

Pruning rarely required.

Uses: specimen or accent, patio tree, in group planting.

ROWAN TREE or EUROPEAN MOUNTAIN ASH

Sorbus aucuparia Zones 3 4 5 6 7 8, plus north area in zone 10

Symmetrical round headed deciduous tree. Fast growing. Height 20′ in 5-10 years, matures to 20-40′; spread equals height. Best in cooler climates.

Flowers: white clusters in spring.

Fruit: orange-red berries in clusters from July. Persist into winter or until taken by birds.

Leaves: pinnate, clear dark green with gray undersides.

FALL COLOR: orange red.

Soil: almost any, well drained; short lived in alkaline conditions. Shallow root system. Mulch, though not touching trunk.

Light: full sun.

Prune young trees if required in winter or early spring.

Uses: shade tree, accent or specimen.

Shade Trees

NORWAY MAPLE *Acer platanoides*

Zones 3 4 5 6 7 8

Large dense round-headed deciduous tree. Very fast growing. Height 20′ in 5-10 years, matures to 70-90′; spreads 40-50′.

Green leaves follow yellow flowers in spring.

FALL COLOR: oranges and yellows, leaves hold well if sheltered from autumn winds.

Soil: moist and well drained. Surface roots take all nourishment from top soil beneath and around tree, and inhibit grass.

Light: full sun or light shade.

TOLERATES urban pollution, coastal climate, and wind.

Prune by thinning branches regularly in late summer and fall. Will bleed excessively if pruned in winter or spring.

Uses: large shade tree.

COLUMNAR NORWAY MAPLE *Acer platanoides 'Columnare'*

Zones 3 4 5 6 7 8

Narrow upright deciduous tree. Very fast growing. Height 20′ in 5-10 years, matures to 50-60′; spreads 25-30′.

Green leaves follow yellow flowers in spring.

FALL COLOR: oranges and yellows, leaves hold well if sheltered from autumn winds.

Soil: moist and well drained. Surface roots take all nourishment from topsoil beneath and around tree.

Light: full sun or light shade.

TOLERATES urban pollution, coastal climate and wind.

Prune by thinning branches in late summer and fall. Will bleed excessively if pruned in winter or spring.

Uses: in restricted areas, accent or specimen, street tree.

CRIMSON KING NORWAY MAPLE *Acer platanoides 'Crimson King'*

Zones 3 4 5 6 7 8

Densely branched deciduous tree. Fast growing. Height 20′ in 8-10 years, matures to 50′; spreads 30-40′.

Leaves retain red purple color throughout summer.

FALL COLOR: intense purple.

Soil: moist and well drained. Surface roots take all nourishment from topsoil beneath and around tree, and inhibit grass.

Light: full sun or light shade.

TOLERATES urban pollution, coastal climate and wind.

Prune by thinning branches regularly in late summer and fall. Will bleed excessively if pruned in winter or spring.

Uses: shade tree, accent or specimen.

SCHWEDLER PURPLE LEAF MAPLE *Acer platanoides 'Schwedleri'* Zones 3 4 5 6 7 8

Densely growing deciduous tree. Fast growth rate. Height 20′ in 8-10 years, matures to 50′; spreads 30-40′.

Leaves open bright red-purple, turning dark purplish green in summer.

FALL COLOR: deep purples and reds.

Soil: moist and well drained. Surface roots take all nourishment from topsoil beneath and around tree, and inhibit grass.

Light: full sun or light shade.

TOLERATES urban pollution, coastal climate, and wind.

Prune by thinning branches regularly in late summer and fall. Will bleed excessively if pruned in winter or spring.

Uses: shade tree, accent or specimen.

RED or SCARLET MAPLE *Acer rubrum*

Zones 3 4 5 6 7 8

Rounded or oval deciduous tree. Very fast growing. Height and spread 20′ in 5-10 years, matures to 50-70′ high, spreads 30-50′.

Flowers red on red twigs, contrasting with gray bark.

Leaves dark green with silvery undersides.

FALL COLOR: orange to deep red.

Soil: moist and well drained. Surface roots take all nourishment from topsoil beneath and around tree.

Light: full sun or light shade.

TOLERATES urban pollution, coastal climate, wind, and high water table.

Prune by thinning branches in late summer and fall. Will bleed excessively if pruned in winter or spring.

Uses: shade tree, accent or specimen.

SILVER or SOFT MAPLE *Acer saccharinum*

Zones 3 4 5 6 7 8 9 10

Large spreading deciduous tree. Very fast growing. Height 20′ in 5-8 years, matures 60-100′; spreads 40-50′. Silver gray bark. Brittle branches which break off in high winds.

Leaves light green with silver undersides.

FALL COLOR: clear yellow-orange or red.

Soil: moist to swampy; surface roots spread to find water, may clog drains.

Light: full sun or light shade.

TOLERATES urban pollution, coastal climate.

Prune by thinning branches in late summer and fall. Will bleed excessively if pruned in winter or spring.

Uses: accent or specimen, shade tree.

SUGAR or HARD MAPLE *Acer saccharum*

Zones 3 4 5 6 7 8

Slender oval headed deciduous tree. Moderate growth rate. Height 20′ in 10-12 years, matures to approx. 75′; spreads 40-50′.

Leaves green in spring and summer.

FALL COLOR: yellow, orange, red.

Soil: moist and well drained. Surface roots take all nourishment from topsoil beneath and around tree.

Intolerant of chemicals, like salt on highways; plant away from roads.

Light: full sun.

Prune by thinning branches in late summer and fall. Will bleed excessively if pruned in winter or spring.

Uses: accent or specimen.

RIVER BIRCH *Betula nigra*

Zones 4 5 6 7 8 9 10

Loose headed upright deciduous tree. Fast growing, single or multiple trunks. Height 20′ in approx. 10 years, matures to 60-70′; single trunk specimen spreads to half its height.

Bark characteristic red-brown color, becoming torn and shredded on mature trees.

Leaves oval and pointed, pale green when young becoming mid green with gray undersides.

FALL COLOR: golden yellow.

Soil: moist or wet loam; tolerates acid, thin and sandy conditions. Wide spreading surface root system.

Light: best in full sun.

TOLERATES heat and drought when established.

Prune if needed in summer and fall.

Uses: accent, street tree.

CANOE or PAPER BIRCH *Betula papyrifera*

Zones 3 4 5 6 7 8

Loose headed upright deciduous tree. Fast growing with single or multiple trunks. Height 20′ in approx. 10 years, matures to 60-70′; single trunk specimen spreads to half its height.

Bark characteristic white and peeling after 4-5 years.

Leaves: oval and pointed, light yellow-green when young becoming mid green with lighter undersides.

FALL COLOR: golden yellow.

Soil: moist loam; tolerates acid, thin and sandy conditions. Wide spreading surface root system.

Light: best in full sun.

Prune if needed, in summer and fall.

Uses: accent, in group plantings.

EUROPEAN WHITE BIRCH *Betula pendula (B. alba, B. verrucosa)* Zones 3 4 5 6 7 8 9 10

Loose headed deciduous tree with pendulous branches. Fast growing with single or multiple trunks. Height 20′ in approx. 10 years, matures to 30-60′; single trunk specimen spreads 20-30′.

Bark characteristic white with black ridges after 4-5 years.

Leaves diamond shaped, light green with lighter undersides.

FALL COLOR: golden yellow.

Soil: wet or dry loam; tolerates acid, thin and sandy conditions. Wide spreading surface root system.

Light: best in full sun.

TOLERATES urban pollution.

Prune if needed in summer and fall.

Uses: accent, in group plantings.

CUT-LEAF EUROPEAN BIRCH *Betula pendula 'Dalecarlica'*

Zones 3 4 5 6 7 8 9 10

Pendulous weeping form when mature. Fast growing. Height 20′ in approx. 15 years, matures to 30-60′; single trunk specimen spreads 20-30′.

Bark characteristic white with black ridges after 4-5 years.

Leaves deeply cut or lobed, light green color; cast light and gentle shade.

FALL COLOR: golden yellow.

Soil: wet or dry loam; tolerates acid, thin and sandy conditions. Wide spreading surface root system.

Light: best in full sun.

TOLERATES urban pollution.

Prune if needed in summer and fall.

Uses: accent.

CLUMP BIRCH

Zones 3 4 5 6 7 8 9 10

Multiple trunk forms of Paper, European White, or River Birch.

All fast growing, with spread wider than that of single trunk specimens.

Bark characteristic of species.

Leaves light to mid green with lighter undersides.

FALL COLOR: golden yellow.

Uses: accent for larger area than single trunk specinen. Shade tree, casting gentle shade.

HACKBERRY or NETTLE TREE *Celtis occidentalis*

Zones 3 4 5 6 7 8 9 10

Wide spreading deciduous tree. Moderate growth rate. Height 25-35′ in 15-20 years, matures to 50-60′. Bark often has characteristic wart-like bumps. Easily transplanted and established.

Leaves oval and pointed; mid to dark green color turning yellow in fall.

Soil: almost any, provided it is deep; tolerates acid or alkaline, wet or dry conditions. Deep root system.

Light: full sun

TOLERATES drought, poor soil, urban conditions; resists wind damage.

Pruning rarely needed.

Uses: accent, street, shade tree.

KATSURA TREE *Cercidiphyllum japonicum*

Zones 4 5 6 7 8 9 10

Conical upright or spreading deciduous tree. Single or multiple trunks. Height 20-25′ in 10-15 years, matures to 40-60′. Single trunk specimen forms upright cone, spreads 30-50′; multiple trunk tree spreads to 40-60′.

Bark characteristic dark brown, shredding and shaggy-looking on mature trees.

Leaves heart shaped; reddish tint in spring turning deep blue green.

FALL COLOR: orange and yellow.

Soil: moist and well drained; tolerates lime.

Light: full sun in northern zones; light shade in south zones 8 and 10, all zone 9. Shelter from cold winds.

Pruning rarely needed.

Uses: accent or specimen, shade tree, casting light and gentle shade.

RUSSIAN OLIVE *Elaeagnus angustifolia*

All zones

Adaptable very fast growing deciduous tree with single or multiple trunks. Height 15-20′ in 5-10 years, matures to 25′; spread equals height. Irregular shape, becoming crooked and gnarled in maturity.

Flowers in early summer, fragrant blossoms, silvery yellow color.

Fruits yellow, eaten by birds; turn grayish color in fall.

Leaves 1½-3″ long, silvery gray.

Soil: almost any, prefers sandy conditions.

Light: full sun.

TOLERATES wind, drought, urban pollution, and coastal climate.

Prune out dead branches at any time. Shape top in winter or early spring.

Uses: accent or specimen for light, silvery effect. Windbreak (multistemmed), street tree.

GREEN ASH *Fraxinus pennsylvanica*

All zones

Very fast growing upright deciduous tree. Height 25′ in 10-12 years, matures to 50-60′; spreads to 20-30′. Male and female flowers on separate trees.

Fruit (female trees only) winged; quickly germinate to produce many unwanted seedlings near tree.

Plant male tree or seedless variety (eg. Marshall's Seedless).

Leaves: mid green, pinnate and fernlike; cast filtered shade that does not inhibit grass.

FALL COLOR: yellows and purples; fallen leaves quickly disintegrate, raking hardly necessary.

Soil: almost any, well drained. Deep root system.

Light: full sun.

Uses: shade tree, accent, street tree.

GINKGO or MAIDENHAIR TREE *Ginkgo biloba*

Zones 4 5 6 7 8 9 10

Slow growing deciduous tree with asymmetrical vertical branches. Does not reach mature form for 20 years; an 8-10′ tree grows to 20′ in 12-18 years, matures to 50-80′ with spread of 40′. Plant male tree only since female fruits are unpleasantly scented.

Leaves fanshaped, 2½-3½″ across; smooth green and leathery texture, free of insect pests and diseases.

FALL COLOR: brilliant yellow.

Soil: almost any, well drained.

Light: best in full sun.

Prune young trees in early spring.

Uses: specimen or accent, street tree.

THORNLESS HONEY LOCUST *Gleditsia triacanthos var. inermis*

Zones 4 5 6 7 8 9 10

Spreading open headed deciduous tree. Very fast growing. Height 25′ in 10-12 years, matures to 35-70′; spread equals height. Easily established. Few if any seed pods.

Leaves 6-8″ long, pinnate and fernlike; unfold late in spring. Dark green color, cast gentle shade.

FALL COLOR: pale yellow; fallen leaves quickly disintegrate, raking rarely necessary.

Soil: deep, fertile, well drained; tolerates alkaline conditions.

Light: full sun.

TOLERATES urban pollution; once established, will tolerate drought.

Prune in fall; older trees rarely need pruning.

Uses: shade tree, accent or specimen, street tree.

SUNBURST HONEY LOCUST *Gleditsia triacanthos var. inermis 'Sunburst'* Zones 4 5 6 7 8 9 10

Pyramidal deciduous tree. Fast growing. Height 25′ in 10-12 years, matures to 35-70′; spreads to only half its height. Easily established.

Leaves 6-8″ long, pinnate and fernlike; unfold in late spring. Golden yellow color until midsummer, then mid to dark green.

FALL COLOR: pale yellow; fallen leaves quickly disintegrate, raking rarely needed.

Soil: deep, fertile, well drained; tolerates alkaline conditions.

Light: full sun.

TOLERATES urban pollution; once established, will tolerate drought.

Prune in fall; older trees rarely need pruning.

Uses: accent tree, especially useful for restricted areas, street tree.

BLACK WALNUT *Juglans nigra*

Zones 4 5 6 7 8

Round headed deciduous tree. Fast growing. Height 20′ in 10-12 years, matures 60-80′. Hard to establish young trees greater than 4-6′ tall. Starts bearing nuts 4-8 years after planting; heavier crops with 2 or more trees in same area, for cross pollination.

Fruit: edible nuts.

Leaves pinnate, mid green color.

FALL COLOR: brownish yellow.

Soil: moist, rich and well drained. Deep root system.

Light: full sun.

TOLERATES urban pollution.

Prune in summer and fall when sap flowing slowly.

Uses: street tree, nut production, wood, shade tree.

SWEET GUM *Liquidambar styraciflua*

Zones 6 7 8 9 10

Cone shaped upright deciduous tree. Young trees attain 25-30′ height in 8-12 years in moist soil, or in 10-15 years in dry conditions. Matures to 60-70′. Hard to transplant trees of 15′ or taller.

Bark silvery brown and furrowed; twigs and young branches winged.

Flowers inconspicuous greenish clusters, followed by prickly seed balls which drop with the leaves.

Leaves open in late spring, hold late in fall; 4-7″ across, dark green.

FALL COLOR: yellows, oranges and reds, hold to late fall.

Soil: moist and deep; add peatmoss or leafmold; tolerates wet conditions. Deep root system.

Light: full sun or light shade.

TOLERATES coastal climate when sheltered from winds.

Prune during winter.

Uses: shade tree, street tree, accent or specimen.

TULIP TREE or TULIP POPLAR *Liriodendron tulipifera*

Zones 5 6 7 8 9 10

Large erect deciduous tree. Very fast growing. Height 25-30′ in 10-12 years, matures to 70-90′; spreads 35-40′. Starts flowering when 10-20 years old.

Flowers: tulip shaped, yellow with orange marking inside petals; fragrant (though high up on tree).

Fruit: 2-3″ light brown seed capsules, persist into winter.

Leaves large and irregular, smooth edged. Light green color turning deeper in summer, glossy.

FALL COLOR: clear golden yellow, flecked brown.

Soil: deep, moist, well drained. Deep root system.

Light: full sun.

TOLERATES coastal climate when sheltered from wind.

Prune young trees in winter.

Uses: shade tree, accent or specimen.

FRUITLESS MULBERRY *Morus alba—Fruitless*

Zones 6 7 8 9 10

Fast growing round topped deciduous tree. A 5-6′ tree grows to 20′ or more with equal spread in 5-10 years. Matures to 50′. Horizontal branches bend lower as they elongate and mature, may require removal.

Leaves 5-6″ long, irregular lobing.

Leaves 3-5″ long, irregular lobing, maple-like.

Soil: great variety, well drained; deep root system.

Light: full sun.

TOLERATES drought once established

Prune in winter.

Uses: fast growing shade tree.

AMERICAN PLANE or SYCAMORE *Platanus occidentalis*

Zones 4 5 6 7 8 9 10

Broad open headed deciduous tree. Very fast growing. Height 20′ in 5-8 years, matures to 40-100′; spread equals height.

Bark characteristic gray brown, flaking off to expose pale greenish yellow patches eventually chalky white.

Flowers inconspicuous, followed by ball shaped seed pods which persist after leaf fall.

Leaves shaped like maple (except alternately arranged), 4-8″ across, dark green all season.

Soil: almost any, prefers deep and moist; tolerates alkaline conditions. Deep root system. Water deeply in dry periods while young; fertilize every one or two years.

Light: full sun or light shade.

VERY TOLERATE of urban pollution; tolerates high water table.

Prune in winter.

Uses: shade tree, accent or specimen, street tree.

LOMBARDY POPLAR *Populus nigra 'Italica'*

All zones

Very fast growing upright columnar deciduous tree with erect close branches. Grows 3-5′ each year when young; an 8-10′ tree grows to 20′ in 3-5 years. Spreads 5-10′. Short lived, very susceptible to canker. Use for short term shelter belt or windbreak. Plant female trees only.

Leaves large dark green.

FALL COLOR: golden yellow.

Soil: almost any, deep and well drained. Roots tend to plug up drains, heave up sidewalks and pavement if too near.

Light: best in full sun.

TOLERATES drought.

Prune in summer and fall.

Uses: quick growing temporary screen for maximum time of 25-30 years.

PIN OAK *Quercus palustris*

Zones 4 5 6 7 8 9 10

Symmetrical pyramid, deciduous tree. Fast growing. Height 20-25′ in 10-20 years, matures to about 75′; spreads 25-40′. Slender branches are spreading or pendulous, and lower ones are retained when planted in the open. Easily established. Acorns take 2 years to mature.

Foliage deeply cut, glossy dark green.

FALL COLOR: crimson.

Soil: moist and rich, prefers acid (pH 5.5-6.5); tolerates dry, rocky and sandy conditions. Use iron chelate or sequestrene in alkaline areas to improve growth and leaf color.

Light: full sun.

TOLERATES urban pollution.

Prune young trees in winter or early spring. Remove lower branches for headroom as needed.

Uses: shade tree, accent, street tree.

PIN OAK *Quercus palustris*

Zones 4 5 6 7 8 9 10

Symmetrical pyramid, deciduous tree. Fast growing. Height 20-25′ in 10-20 years, matures to about 75′; spreads 25-40′. Slender branches are spreading or pendulous, and lower ones are retained when planted in the open. Easily established. Acorns take 2 years to mature.

Foliage deeply cut, glossy dark green. Holds dry leaves all winter.
FALL COLOR: crimson.

Soil: moist and rich, prefers acid (pH 5.5-6.5); tolerates dry, rocky and sandy conditions. Use iron chelate or sequestrene in alkaline areas to improve growth and leaf color.

Light: full sun.

TOLERATES urban pollution.

Prune young trees in winter or early spring. Remove lower branches for headroom as needed.

Uses: shade tree, accent, street tree.

DRAGON-CLAW WILLOW *Salix matsudana 'Tortuosa'*

Zones 4 5 6 7 8 9 10

Slow growing erect deciduous tree with twisted branches and leaves. Multiple or single trunks. Matures to about 30′, spreads 20-30′.

Bark yellow-gray, young branches and twigs yellow to olive green. Branches brittle and break off in wind.

Flowers inconspicuous, ¾″ green catkins with first leaves in spring.

Leaves long and narrow, light green with silvery reverse; twisted.

FALL COLOR: yellow.

Soil: moist or wet all year. Water well if soil dries out in summer. Shallow root system inhibits growth of grass.

Light: full sun.

Prune in summer and fall; take out lower branches for headroom.

Uses: specimen or accent, shade tree.

WEEPING WILLOW *Salix babylonica*

Zones 4 5 6 7 8 9 10

Broad headed deciduous tree with pendulous flexible weeping branches. Very fast growing. Height 20′ in 3-6 years, matures to 50-60′; spread equals height.

Bark pale gray-green, young branches and twigs yellow to olive green. Branches brittle and break off in wind.

Flowers inconspicuous, 1″ long catkins borne before leaves fully expanded in spring.

Leaves long and narrow, light green with silvery reverse.

FALL COLOR: yellow.

Soil: moist or wet all year. Water well if soil dries out in summer. Shallow root system prevents even grass growing well under tree.

Light: full sun.

Prune lower branches in summer and fall, for headroom and to allow tree to weep attractively.

Uses: specimen or accent, shade tree.

CRIMEAN LINDEN *Tilia ×euchlora*

Zones 4 5 6 7 8 and north in zone 10

Densely conical deciduous tree. Fast growing. Height 20′ in 8-10 years, matures to 40-60′; spreads to half its height.

Flowers in midsummer, yellow and fragrant, attracts bees. Followed by tiny inedible fruits which persist into winter.

Leaves medium length, glossy and dark green.

FALL COLOR: yellow with cold nights, in mild climates leaves remain green until killed by frost.

Soil: deep, moist, well drained. Deep root system.

Light: full sun or light shade.

TOLERATES urban pollution.

Pruning rarely needed.

Uses: shade tree.

LITTLE LEAF LINDEN *Tilia cordata*

Zones 3 4 5 6 7 8 and north in zone 10

Symmetrical pyramid shaped deciduous tree. Fast growing. Height 20′ in 10-12 years, matures to 50-70′; spreads 25-30′.

Flowers in midsummer, yellow-white and very fragrant, attractive to bees. Followed by tiny inedible fruit which persist into winter.

Leaves 1½-3″ long, lopsided, mid green.

FALL COLOR: yellow with cool nights; in mild climates, leaves remain green until killed by frost.

Soil: deep, moist, well drained. Deep root system.

Light: full sun or light shade.

TOLERATES urban pollution.

Pruning rarely needed. Remove suckers at ground level.

Uses: street tree, shade tree.

CHINESE ELM *Ulmus parvifolia 'Drake'*

Zones 6 7 8 9 10

Rapid growing deciduous tree, spreading with pendulous branches. An 8-10′ tree grows to 20′ in 10-12 years, matures to 40-50′; spread equal to height.

Leaves 1-2″ long, glossy mid to dark green.

FALL COLOR: green to purplish red.

Soil: almost any, deep and well drained; shallow root system.

Light: full sun or light shade.

Prune in fall if needed.

Uses: shade tree.

Food-Producing Plants

DWARF APPLES *Malus hybrid*

Zones 4 5 6 7 8 (check varieties for hardiness in your area)

Deciduous fruit tree reaching 6-8′ tall. Variety grafted onto dwarfing root stock. Starts flowering and bearing fruit 2-3 years after planting. Requires different apple variety within 50′ for adequate cross pollination to yield fruit. Life span 30-50 years.

Flowers in early spring; fragrant white blossoms in clusters.

Fruit ripens in summer or fall depending on variety.

Soil: Prefers slightly acid reaction. (pH 5.5 to 6.5) Add ⅓ volume peat moss or leaf mold when planting. Plant tree at same level or 1-2″ deeper than in nursery. Stake as necessary. Leave depression to form watering saucer in first summer; Mound soil up for winter to prevent frost damage and heaving.

Position fruit tree on slope so that cold air flows away and reduces danger of frost damage.

Light: Best in full sun.

Prune: Young tree from 3-5 strong well angled branches (angle greater than 45° with trunk); cut these back to 6-10″. Each subsequent spring prune to develop open pyramidal form or to train against support; avoid dense, over long and crossed branches. Once tree starts bearing fruit, light pruning only is needed.

Uses: Free standing tree, space 10-12′ apart in groups. Container free for patios, pruned to form hedge, esparieled against wall or fence.

DWARF APPLES *Malus hybrids*

Deciduous fruit tree reaching 6-8′ tall. Variety grafted onto dwarfing root stock. Starts flowering and bearing fruit 2-3 years after planting. Requires different apple variety within 50′ for adequate cross pollination to yield fruit. Life span 30-50 years.

Flowers in early spring; fragrant white blossoms in clusters.

Fruit ripens in summer or fall depending on variety.

Soil: Prefers slightly acid reaction. (pH 5.5 to 6.5) Add ⅓ volume peat moss or leaf mold when planting. Plant tree at same level or 1-2″ deeper than in nursery. Stake as necessary. Leave depression to form watering saucer in first summer; Mound soil up for winter to prevent frost damage and heaving.

Zones 4 5 6 7 8 9 (check cultivars for hardiness in your area)
Position fruit tree on slope so that cold air flows away and reduces danger of frost damage.

Light: Best in full sun.

Prune: Young tree from 3-5 strong well angled branches (angle greater than 45° with trunk); cut these back to 6-10″. Each subsequent spring prune to develop open pyramidal form or to train against support; avoid dense, over long and crossed branches. Once tree starts bearing fruit, light pruning only is needed.

Uses: Free standing tree, space 10-12′ apart in groups. Container free for patios, pruned to form hedge, esparieled against wall or fence.

DWARF APPLE (Several cultivars on one tree)

Zones 4 5 6 7 8 (cultivars suited to different areas)

Deciduous fruit tree reaching 6-8′ tall. More than one variety grafted onto dwarfing rootstock; cross pollination assures fruit crop without additional trees. Starts flowering and bearing fruit 1-2 years after planting. Life span 20-30 years.

Flowers in early spring; fragrant white blossoms in clusters.

Fruit ripen in summer or fall, depending on varieties.

Soil: prefers slightly acid reaction (pH 5.5-6.5). Add ⅓ volume peatmoss, compost or leafmold when planting. Plant tree at same level or 1-2″ deeper than in nursery. Stake as necessary. Leave depression to form watering saucer in first summer; mound soil up for winter to prevent frost damage and heaving.

Position fruit tree on slope so that cold air flows away and reduces danger of frost damage: or place where tree is protected from frost particularly when new spring growth begins.

Light: best in full sun.

Prune carefully to avoid cutting out grafts. Different varieties may require individual treatment; each spring reduce vigorous vegetative growth and prune to avoid dense, long, or crossed branches. Once fruiting starts, only light pruning will be needed.

Uses: container tree for patio use, in small gardens.

BLACKBERRIES (Boysenberry, Loganberry, Youngberry) *Rubus* All zones (cultivars suited to different areas)

Deciduous, usually thorny shrubs with stiffly erect or trailing stems (canes). Height and spread controlled with pruning and training. Productive life about 20 years.

Flowers: clusters of single white or pinkish rose-like blossoms open after leaves.

Fruit: 1-2″ black or red edible fruits borne on 2nd year canes.

Soil: fertile, moist and well drained, prefers slightly acid to neutral reaction (pH 5.5-7.5). Add compost, peatmoss or leafmold. Mulch in summer with 2-3″ layer of wood chips, sawdust or ground bark.

Light: best in full sun.

Prune newly planted canes to 6-8″. Tip the current season's new canes to promote branching. In spring reduce total number of canes and tip the side branches. Prune out fruited canes in fall or spring.

Fertilize in early spring.

Uses: fruiting hedge or barrier: plant **bush** type berries 4-5′ apart, thin new canes to space about 10″ apart and tip them at 3′. Space rows about 6′ apart.
Hedge (with supports): plant **vine** type berries (less hardy than the bush types) about 6′ apart, tip new canes when they reach 8′ in length and tie them in bundles of 4, to horizontal wires. Thin canes to about 16 per plant.

BLUEBERRIES (Highbush, Rabbiteye, Lowbush) *Vaccinium*

All zones (cultivars suited to different areas)

Deciduous fruiting shrubs, some evergreen in the south. Upright vase shape bushes or low spreading habit. All require another variety for adequate cross pollination for fruit production. Productive life about 40 years.

Flowers in spring, clusters of small white blossoms.

Leaves small and dark green.
FALL COLOR: brilliant reds and scarlets.

Fruit ripen in mid to late summer, edible blue berries.

Soil: very moist and acid (pH 5.0). Prepare an acid bed if necessary by replacing top 18-24″ of soil with equal volumes of acid peatmoss, leafmold and coarse sand; or plant into sunken containers containing same acid soil mix. Iron chelate or sequestrene will improve leaf color. Mulch permanently with 4-8″ layer of pine needles, oak leaves, peatmoss, wood chips or (add fertilizer first) sawdust. Do not cultivate around bushes.

Light: best in full sun.

Prune newly planted blueberries to half height. Each spring remove thin growth and old wood before buds start to swell. Build strong plants by picking off flowers to prevent fruiting for first 2-3 years.

Fertilize young plants with weak mixture to avoid chemical damage to roots.

Uses: massed planting for screen—**Highbush** (north) planted 5′ apart in rows spaced at about 8′. **Rabbiteye** (south) planted 8′ apart in rows spaced at 8-10′.
ground cover—**Lowbush** blueberries planted 2-4′ apart grow about 8″ high.

SWEET CHERRIES *Prunus avium*

Zones 5 6 7 8

Deciduous fruiting tree with variety budded onto vigorous rootstock. Height reaches about 25′ with equal spread. Sweet cherries require cross-pollination for fruit production; more than one variety needed.

Flowers in early spring before leaves unfurl. Single white blossoms followed by edible fruits ripening in midsummer.

Soil: moist and well drained, some lime preferred (pH 6.0-8.0). Add compost, leafmold or peatmoss in volume equal to soil when planting. Stake newly planted trees until established (2-3 years). Shallow root system; do not cultivate around trees. Mulch after soil has warmed in spring to retain moisture and to add nutrients.

Light: best in full sun.

Prune young tree to 2-4 well spaced and angled branches (angle with trunk more than 45°). Shorten these to 8-10″. After this prune lightly in late summer or when dormant to remove overcrowded or weak growth. Sweet cherries have naturally upright growth and need very little pruning, save that to encourage outer, spreading branches.

Fertilize in early spring if previous season's leaves were pale or yellow.

Uses: free standing tree, plant 30′ apart.

TART CHERRIES *Prunus cerasus*

Zones 4 5 6 7 8

Deciduous fruiting tree with variety budded onto vigorous rootstock. Height reaches 10-15′, spread 12-20′. All tart, or sour, cherries are self-fruitful, producing fruit without needing cross pollination from another variety.

Flowers in early spring before leaves unfurl. Single white blossoms followed by edible fruits ripening in midsummer.

Soil: moist and well drained, some lime preferred (pH 6.0-8.0). Add compost, leafmold or peatmoss in volume equal to soil when planting. Stake newly planted trees until they are established (2-3 years). Shallow root system; do not cultivate around trees. Mulch after soil has warmed in spring to retain moisture and add nutrients.

Light: best in full sun.

Prune young trees to 2-4 well spaced and angled branches (angle with trunk more than 45°). Shorten these to 8-10″. After this prune lightly in late summer or when dormant to remove overcrowded or weak growth. Tart cherries are natural spreaders: prune to encourage upright branches.

Fertilize in early spring if previous season's leaves were pale or yellow.

Uses: free standing trees, plant 25′ apart.

CURRANTS (red and white) *Ribes*

Zones 4 5 6 7 8 9

Deciduous vase-shaped shrubs with height and spread of 3-4′. Productive life 20-30 years.

Leaves mid green; clusters of inconspicuous yellowish flowers.

Fruit: red or white (or black), edible currants borne on shoots (canes) more than one year old.

Soil: heavy, slightly acid (pH 5.5-7.0). Add compost, peatmoss or leafmold.

Light: full sun or light shade.

Prune newly planted currant bushes to 6-8″ above soil. Leave unpruned for 3 seasons, then remove 3 year old canes to ground level and select 6-8 of the strongest current season's canes. Maintain succession of one, two, and three year old canes in this way.

Fertilize with a manure mulch or with dry fertilizer in fall or early spring.

Uses: single bush, or massed to make a fruiting screen or hedge. Plant 5′ apart.

NOTE: currants are alternate hosts for the White Pine Blister Rust disease, and are prohibited in some areas. Check the local Agricultural Extension Service for regulations in your area.

GOOSEBERRIES *Ribes*

Zones 4 5 6 7 8 9

Deciduous thorny bushes, height and spread 2-4′. Productive life about 25 years.

Leaves light green, contrast with light gray-brown bark.

Fruit about ¾″ across; greenish-yellow, pink or red edible berries. Best fruit borne on 2nd and 3rd year canes; ripens in midsummer.

Soil: medium to heavy texture, fertile, slightly acid (pH 5.5-6.5). Add compost, peatmoss or leafmold.

Light: full sun or light shade.

Prune newly planted gooseberries to 6-8″ above soil. After one season prune in winter to retain 6 good canes, cutting everything else to the ground. Next winter remove all new canes except 3. In the third winter cut out 3 of the 6 oldest canes as well as all new ones except 3. Each subsequent winter remove the 3 oldest canes and replace with 3 new ones, to maintain a succession of one, two and three year old canes.

Fertilize with a manure mulch or with dry fertilizer in fall or early spring.

Uses: single bush in shrub border, or massed to make thorny yet fruitful screen or hedge. Plant 4-5′ apart.

NOTE: gooseberries are alternate hosts for the White Pine Blister Rust disease, and are prohibited in some areas. Check the local Agricultural Extension Service for regulations in your area.

DWARF PEACH *Prunus Persica*

Zones 5 6 7 8 9 (excluding southern Florida) and 10.

Deciduous fruit tree reaching height of 7-8′. Variety grafted onto dwarfing root stock. Starts flowering and bearing fruit 2-3 years after planting. Requires different variety within 50′ for adequate cross pollination to yield fruit.
Life span 10-20 years.

Flowers in early spring before leaves unfold; single pink blossoms on previous seasons growth.

Fruit ripens in mid summer.

Soil: Acid to neutral (pH 6.0-7.0). Add peat moss, compost or leaf mold in equal volume with soil when planting.

Position fruit tree on slope so that cold air flows away and reduces danger of frost damage.

Light: Best in full sun.

Prune young tree from 3-5 strong well angled branches (angle greater than 35° with trunk); cut these back 2-4″. Each subsequent spring prune to remove excessive, low hanging or crossed branches. Prune lightly until established and bearing fruit.

Thin Fruit as necessary when fruits start to swell. One fruit for every 35 leaves or at 6-8″ spacing along branches, is the recommended load.

Uses: Free standing tree, space 10-12′ apart in groups. Container free for patios, pruned to form hedge, esparieled against wall or fence.

PEARS *Pyrus communis*

Zones 4 5 6 7 8 (cultivars suited to different areas)

Upright deciduous fruiting tree with one or more varieties grafted or budded onto standard or dwarfing rootstock. Standard size trees reach 15-18′, dwarf trees 8-10′ tall. More than one variety needed to assure cross pollination for fruit production. Starts flowering and bearing fruit 1-2 years after planting. Life span 50-100 years.

Leaves glossy dark green.

Flowers in early spring, white blossoms in clusters.

Fruit ripen in summer or fall, depending on variety. Harvest before ripe to avoid brown cores; pick pears when stem swells and will separate at the swelling when fruit is lifted. Store cool (31-32°F) or in refrigerator (40°) for at least 1-2 weeks before ripening in warm dark room.

Soil: heavy and moist yet well drained; acid to neutral (pH 6.0-7.0). Add ⅓ volume peatmoss, compost or leafmold when planting. Plant tree at same level as in nursery, keeping graft union above soil level. Stake as necessary. Grow in grassed area if possible so pear competes with grass for moisture and nutrients; slower vegetative growth reduces chance of attack by fireblight disease.

Light: best in full sun.

Prune newly planted young tree to 4-8 well angles branches (angle with trunk greater than 45°) plus central, or leader shoot. Shorten branches to half and tip leader. After this prune as little as possible to remove crossed or excessively long shoots, in early spring. Remove diseased branches cutting back 12″ into healthy looking wood.

Uses: free-standing tree on grassy slope or lawn; space standard trees 15′ apart, dwarfs 10′ apart. Good for small areas, with narrow upright growth habit. Espaliered against wall, fence or trellis.

PLUMS (Prunes, Green Gages and Damsons) *Prunus*

Zones 4 5 6 7 8 9 (excluding southern Florida) and 10 (cultivars suited to different areas)

Deciduous trees with spreading or vase shaped growth. Fruiting variety budded or grafted onto standard or dwarfing rootstock. Standard trees reach 15-20′ tall with equal spread, dwarf 8-10′. Start flowering and fruiting 2-3 years after planting. Damsons, Green Gages and some other plum varieties are self fruitful; others require another variety for cross pollination and fruit production.

Flowers in early spring before leaves unfurl, single white or light pink blossoms.

Fruit ripen in midsummer.

Soil: fertile, neutral reaction (pH 6.0-8.0). Add compost, peatmoss or leafmold in volume equal to soil when planting. Stake newly planted trees until they are established.

Light: best in full sun.

Prune young trees to 3-5 well angled branches (angle with trunk more than 45°). Shorten these to 6-8″. After this, prune lightly in late summer or when dormant to thin crowded growth and to remove weak or dead wood.

Thin fruit (not Damsons) to improve size and flavor, reducing crop to one fruit for every 1-3″ shoot length.

Fertilize in early spring each year to maintain vigor and fruit production.

Uses: free standing tree in lawn or border; space standard size trees 20′ apart, dwarfs 10-12′ apart. Dwarf trees good in small areas, for patios.

RASPBERRIES *Rubus*

Zones 3 4 5 6 7 8 (cultivars suited to different areas)

Deciduous thorny shrubs with long erect or rambling shoots (canes). Height and spread controlled with pruning and training. Productive life about 10 years.

Flowers: clusters of single white rose-like blossoms open after leaves.

Fruit borne on 2nd year canes: varieties have red, yellow black or purple fruits.

Common raspberry varieties crop once, in midsummer.

Everbearing varieties crop in early summer and again on the same canes in late summer and fall.

Soil: fertile, moist and well drained, prefers slightly acid reaction (pH 5.5-7.0). Add compost, peatmoss, or leafmold.

NOTE: raspberries are susceptible to soil-borne diseases and are best **not** planted into the same soil where eggplants, peppers, potatoes or tomatoes have been raised within three years.

Prune newly planted canes to 2″ above soil level, and remove these stubs when new growth begins. Tip the current season's new canes to promote branching. In spring reduce total number of canes and shorten them to 3-4′. Remove fruited canes in fall or spring, cutting off at ground level.

Light: best in full sun.

Fertilize in early spring.

Uses: fruiting hedge or screen: plant 2-3′ apart, thin new canes to 3 or 4 per foot. Restrict sucker growth to strip about 1′ wide. Support canes between horizontal wires 12-18″ apart. Space rows 7-8′ apart. Single clumps or hills: space plants 5-7′ apart, thin new canes to 6 to 8 per clump. Support on post.

RHUBARB *Rheum rhaponticum*

All zones

Hardy herbaceous perennial with large spreading leaves and edible leaf stalks.

Leaves dark green with pink or red stalk and veins. Leaf blade 12″ or more across, stalk 18-36″ long. Although stalks are edible, leaf tissue is poisonous.

Soil: highly organic and fertile, acid to neutral (pH 5.5-7.0). Dig out 2′ by 2′ area and replace subsoil with 6-8″ layer of pure compost or manure; mix compost or manure in equal volume with topsoil.

Plant crowns so root tops are 3-4″ below soil surface. Space plants 3′ apart. Let rhubarb grow undisturbed by cultivations. Start harvesting two years after planting; take stalks that are 1″ or more in diameter. Cut off flowering stalks before seeds ripen.

Light: best in shade.

Fertilize in fall when leaves have died back, with 3″ mulch of compost or manure, or apply dry garden fertilizer around plants and water in.

Uses: ground cover between trees and tall shrubs: eliminates need for constant cultivation. To force rhubarb in winter: lift plants and allow to freeze without drying out, for several weeks. Place in container (e.g. wooden box) with soil or sand; moisten well and place in dark place above freezing (optimum 50-65°F) to force.

STRAWBERRIES *Fragaria ×ananassa*

Zones 4 5 6 7 8 9 10 (cultivars suited to different areas)

Herbaceous perennial plants growing 6-8″ tall and spreading about 12″; runners increase spread with new plants. Edible fruits: productive life span 1-3 years.

Fruit borne on buds developed in previous late summer or during current growing season.

June bearing strawberry varieties crop once, in early to mid-summer in most areas; in coastal California they crop continuously from April; in Florida fruit often ripens in midwinter.

Everbearing strawberries crop once in early summer, then after a brief pause fruit ripens continuously until frost.

Soil: fertile, well drained and slightly acid (pH 5.5-6.5). Add compost, leafmold, peatmoss or manure. Water well especially when fruiting. Mulch in fall when nights average 20°F or lower, using 3-5″ layer of straw or salt marsh hay. Remove most of the mulch in spring when new growth begins.

NOTE: strawberries are susceptible to soil borne diseases and are best not planted into the same soil where eggplant, potatoes, tomatoes, melons, okra, cotton or raspberries have been raised within three years.

Light: best in full sun.

Fertilize newly planted strawberries with a half strength liquid fertilizer. Add dry mixture to soil (not leaves) in late summer and water in.

Spray once in spring as soon as growth begins, using a multipurpose mixture at the recommended rate.

Uses: ground cover: space 12-18″ apart in all directions; pinch off flowers early in first season to strengthen plants. Remove runners to contain spread. Edging for borders: plant in single row 12-24″ apart. Container plant for patios, small gardens, window boxes.

Ground Covers

BUGLEWEED or CARPET BUGLE *Ajuga reptans*

All zones

Low growing and spreading herbaceous perennial; develops mounds up to 6″ high, spreads rapidly by stolons (overground stems). Easy to multiply by division.

Flowers in May and June, compact spikes about 6″ tall. Varieties in blue, white, pink and purple.

Leaves about 4″ long, cultivars may be deep green, purple, bronze or variegated. Fall color a reddish bronze. Leaves may be held right through winter.

Soil: almost any, well drained and fertile. Damp humid conditions tend to encourage rot; maintain active healthy growth with regular fertilization.

Light: sun or shade.

Uses: ground cover, plant 6-12″ apart, edging for borders and along walks.

PURPLE WINTERCREEPER *Euonymus fortunei 'Colorata'* Zones 4 5 6 7 8 9 10

Evergreen spreading creeper, erect stems to height of about 4′ and spreading 4′ in 8-10 years.

Leaves: 1″ ovals, dark green turning reddish purple in fall. Holds leaves through all but severest winter.

Soil: almost any, wet or dry.

Light: full sun or shade.

Prune at any time to shape. Responds well to shearing.

Uses: low hedge, ground cover, foundation planting.

ENGLISH IVY *Hedera helix*

All zones

Evergreen woody creeper.

Leaves dark green and glossy, characteristic ivy shape. Veins often marked with silver.

Soil: any.

Light: best in shade.

Prune in late winter, cutting back close to ground or support. Trim off excessive growth and thin out matted stems in summer.

Uses: ground cover, banks, walls, trellis, screen.

SHORE JUNIPER *Juniperus conferta*

Zones 6 7 8 9 10

Slow growing trailing juniper, rarely exceeds 12″ high. Spreads 6-8′ in 20-25 years.

Foliage bright green, soft texture.

Soil: almost any, well drained; tolerates dry, sandy, acid and alkaline conditions.

Light: best in full sun.

TOLERATES coastal climate.

Pruning: with hand clippers to remove unwanted shoots.

Uses: ground cover, & on gentle slopes and banks, rock gardens, patio, foundation planting.

WAUKEGAN JUNIPER *Juniperus horizontalis 'Douglasii'*

Zones 3 4 5 6 7 8 9 10

Moderate growth, spreading, branches nearly horizontal. Spreads approx. 10′ in 10-15 years, matures 12-16′ across. Height 1-1½′.

Foliage blue green, turns purplish in winter.

Soil: almost any, well drained; tolerates dry, rocky, acid and alkaline conditions.

Light: best in full sun.

Pruning: rarely needed.

Uses: rock gardens, ground cover, cover for walls, stones, gentle slopes and banks, patio.

ANDORRA JUNIPER *Juniperus horizontalis 'Plumosa'*

Zones 3 4 5 6 7 8 9 10

Moderate growth, spreading, with trailing, semi-erect branches. Spreads 6-8′ in 10-15 years, matures 10-15′ across. Height 1-3′.

Foliage silver green in spring and summer, turns pinkish to silvery purple in fall and winter.

Soil: almost any, well drained; tolerates dry, rocky, acid and alkaline conditions.

Light: best in full sun.

Pruning: clip or shear during growing season to maintain desired size.

Prune to encourage dense growth.

Uses: rock gardens, foundation planting, ground cover, and on gentle slopes and banks, patio.

COMPACT YOUNGSTOWN ANDORRA JUNIPER *Juniperus horizontalis 'Plumosa Compacta Youngstown'*

Zones 3 4 5 6 7 8 9 10

Moderate growth, spreading with trailing, semi-erect branches. Spreads 6-8′ in 10-15 years, matures 10-15′ across. Height 10-20″.

Foliage silvery green, color holds all year.

Soil: almost any, well-drained; tolerates dry, rocky, acid and alkaline conditions.

Light: best in full sun.

Pruning: clip or shear during growing season as required.

Uses: rock gardens, foundation planting, ground cover on gentle slopes and banks, patio.

BLUE RUG JUNIPER *Juniperus horizontalis 'Wiltonii'*

Zones 3 4 5 6 7 8 9 10

Spreading with horizontal branches, moderate growth rate. Spreads to 10′ in 10-15 years, matures to 12-16′ across; height approx. 6″.

Foliage strong silvery blue color all year.

Soil: almost any, well drained; tolerates dry, rocky, acid and alkaline conditions.

Light: best in full sun.

Pruning: rarely needed.

Uses: rock gardens, ground cover, cover for walls, stones, gentle slopes and banks, patio.

DWARF JAPANESE GARDEN JUNIPER *Juniperus chinensis 'Procumbens Nana'* Zones 4 5 6 7 8 9 10

Slow growth, spreading, horizontal branches. Spreads 5′ in 10-15 years, matures 10-15′ across. Height approx. 6″.

Foliage blue green, turns purplish in cold weather.

Soil: almost any, well drained: tolerates dry, rocky, acid and alkaline conditions.

Light: best in full sun.

Pruning: rarely needed.

Uses: rock gardens, ground cover, and on gentle slopes and banks, patio. Good bonsai plant.

VARIEGATED JAPANESE GARDEN JUNIPER *Juniperus chinensis 'Procumbens Variegata'*

Zones 4 5 6 7 8 9 10

Slow growing, spreading with horizontal branches. Spreads 5′ in 10-15 years, matures 10-15′ across. Height approx. 12″.

Foliage gray green highlighted with creamy white coloring.

Soil: almost any, well drained; tolerates dry, rocky, acid and alkaline conditions.

Light: best in full sun.

Pruning: rarely needed.

Uses: rock gardens, ground cover, & on gentle slopes and banks, patio. Good bonsai plant.

SCANDIA JUNIPER *Juniperus sabina 'Scandia'*

Zones 4 5 6 7 8 9 10

Moderate to slow growth, dense spreading habit, with semi-erect branches. Spreads 5-7′ in 10-15 years, matures 9-12′ across. Height 2-4′.

Foliage dark green and feathery, holds color all year.

Soil: almost any, well drained; tolerates dry, rocky, acid and alkaline conditions.

Light: best in full sun.

TOLERANT of poor soils and city conditions.

Pruning: clip or shear during growing season to maintain desired size.

Uses: rock garden, foundation planting, ground cover, patio.

TAM JUNIPER *Juniperus sabina 'Tamariscifolia'*

Zones 4 5 6 7 8 9 10

Moderate to slow growth, dense mound-like spreading habit with semi-erect branches. Spreads 4-5′ in 10-15 years, matures 9-12′ across. Height 2-4′.

Foliage blue green, thick and feathery, holds color all year.

Soil: almost any, well-drained; tolerates dry, rocky, acid and alkaline conditions.

Light: best in full sun.

TOLERATES poor soils and urban pollution.

Pruning: clip or shear during growing season to maintain desired size.

Uses: rock garden, foundation planting, ground cover, patio, overhanging walls.

JAPANESE SPURGE *Pachysandra terminalis*

All zones

Evergreen spreading subshrub, 6-8″ tall.

Flowers in spring, tiny spikes of white petal-less blossoms, tinged purple.

Foliage: glossy and leathery, dark green.

Soil: almost any, moist and fertile, well drained.

Light: best in shade.

Uses: ground cover, rock garden.

LAVENDER COTTON *Santolina chamaecyparissus*

Zones 7 8 9 10

Shrublike perennial, grown as annual in all zones. Unpruned height as perennial reaches 2′, spreads 3′. Pinch off flower buds to encourage branching and dense growth.

Flowers: ¾″ button-like yellow blossoms from early summer.

Foliage needle-like, clothes stems, light gray or blue-green color all year in milder climates.

Soil: almost any, well drained.

Light: best in full sun.

Prune in spring, cutting back to 4-6″. Shear regularly to keep growth bushy, and pick off flower buds for more branching.

Uses: edging and low hedge, foundation planting.

GREEN SANTOLINA *Santolina virens*

Zones 6 7 8 9 10

Shrublike perennial, grown as annual in all zones. Unpruned height as perennial reaches 2′, spreads 3′. Pick off flower buds to encourage branching and dense growth.

Flowers: ¾″ button-like yellow blossoms from early summer.

Foliage needle-like, clothes stem, light green color all year in milder climates.

Soil: almost any, well drained.

Light: best in full sun.

Prune in spring, cutting back to 4-6″. Shear regularly to keep growth bushy, and pick off flower buds for more branching.

Uses: edging and low hedge, foundation planting.

TRAILING PERIWINKLE or T. MYRTLE *Vinca minor*

Zones 4 5 6 7 8 9 10

Low growing evergreen creeper with trailing stems. Height to 6″, spreads rapidly forming dense ground cover.

Flowers in early spring; lavender blue blossoms about 1″ across. Varieties also in white, purple or blue.

Leaves ½-¾″ glossy dark green.

Soil: Deep and moist but not wet. Add peat moss or leaf mold to improve water retention.

Light: or deep shade.

Uses: Ground cover for shady areas, plant 12-18″ apart, background to bulb flowers, window boxes, patio containers.

TRAILING PERIWINKLE or T. MYRTLE *Vinca minor*

Zones 4 5 6 7 8 9 10

Low growing evergreen creeper with trailing stems. Height to 6″, spreads rapidly forming dense ground cover.

Flowers in early spring; lavender blue blossoms about 1″ across. Varieties also in white, purple or blue.

Leaves ½-¾″ glossy dark green.

Soil: Deep and moist but not wet. Add peat moss or leaf mold to improve water retention.

Light: or deep shade.

Uses: Ground cover for shady areas, plant 12-18″ apart, background to bulb flowers, window boxes, patio containers.

Roses

ROSES, HYBRID TEA

All zones

Deciduous shrubs with many variations in size, form, and flower form and color. Flowers throughout summer or with one main flush in early summer. Fruits often decorative orange or red berries (hips). Leaves with 3 or 5 leaflets, glossy mid to dark green; 5, 7 or 9 leaflets on species roses.

Soil: almost any, well drained; add peatmoss or leafmold. Prefers slightly acid conditions (pH 5.5-7.0). Mulch in spring when new growth begins to retain moisture.

Light: best in full sun.

Prune in early spring as buds start growing, to bring light and air to center of plant. Remove all dead wood.

HYBRID TEA ROSES: select up to 7 strong basal shoots, prune each one back to 12-15″ with top bud to the outside.

ROSES, FLORIBUNDA

All zones

Deciduous shrubs with many variations in size, form, and flower form and color. Flowers throughout summer or with one main flush in early summer. Fruits often decorative orange or red berries (hips). Leaves with 3 or 5 leaflets, glossy mid to dark green; 5, 7 or 9 leaflets on species roses.

Soil: almost any, well drained; add peatmoss or leafmold. Prefers slightly acid conditions (pH 5.5-7.0). Mulch in spring when new growth begins to retain moisture.

Light: best in full sun.

Prune in early spring as buds start growing, to bring light and air to center of plant. Remove all dead wood.

FLORIBUNDA ROSE BUSHES: select up to 7 strong basal shoots, prune each one back to 12-15″ with top bud to the outside.

ROSES, CLIMBING

All zones

Deciduous shrubs with many variations in size, form, and flower form and color. Flowers throughout summer or with one main flush in early summer. Fruits often decorative orange or red berries (hips). Leaves with 3 or 5 leaflets, glossy mid to dark green; 5, 7 or 9 leaflets on species roses.

Soil: almost any, well drained; add peatmoss or leafmold. Prefers slightly acid conditions (pH 5.5-7.0).

Mulch in spring when new growth begins to retain moisture.

Light: best in full sun.

Prune in early spring as buds start growing, to bring light and air to center of plant. Remove all dead wood.

CLIMBING ROSES: flowers develop at shoot tips; prune all shoots to encourage branching. Tie in main stems.

ROSES, RAMBLING

All zones

Deciduous shrubs with many variations in size, form, and flower form and color. Flowers throughout summer or with one main flush in early summer. Fruits often decorative orange or red berries (hips). Leaves with 3 or 5 leaflets, glossy mid to dark green; 5, 7 or 9 leaflets on species roses.

Soil: almost any, well drained; add peatmoss or leafmold. Prefers slightly acid conditions (pH 5.5-7.0). Mulch in spring when new growth begins to retain moisture.

Light: best in full sun.

Prune in early spring as buds start growing, to bring light and air to center of plant. Remove all dead wood.

RAMBLING ROSES: select up to 7 strong basal shoots, prune each one back to 12-15″ with top bud to the outside.

ROSES, SHRUB, PILLAR AND SPECIES

All zones

Deciduous shrubs with many variations in size, form, and flower form and color. Flowers throughout summer or with one main flush in early summer. Fruits often decorative orange or red berries (hips). Leaves with 3 or 5 leaflets, glossy mid to dark green; 5, 7 or 9 leaflets on species roses.

Soil: almost any, well drained; add peatmoss or leafmold. Prefers slightly acid conditions (pH 5.5-7.0). Mulch in spring when new growth begins to retain moisture.

Light: best in full sun.

Prune in early spring as buds start growing, to bring light and air to center of plant. Remove all dead wood.

SHRUB, PILLAR AND SPECIES ROSES: select up to 7 strong basal shoots, prune each one back to 12-15″ with top bud to the outside.

Hedges and Screens

JAPANESE BOXWOOD *Buxus microphylla var. japonica*

Zones 7 8 9 10

Evergreen shrub growing to 6′ if unpruned.

Flowers in spring, inconspicuous creamy clusters at ends of branches.

Leaves ½-1″ ovals, glossy green, retained through winter.

Soil: almost any, moist and well drained. Mulch with 1″ layer of coarse peatmoss, wood chips or peanut shells to retain moisture in summer. Syringe during dry spells until established (1-2 years).

Light: full sun or light shade. Protect from winter winds using burlap or other screening.

Prune only to remove dead branches.

Uses: formal or informal hedges; plant at 15-30″ and upper third of all growing shoots can be removed to encourage bushy growth.

THORNY ELAEAGNUS *Elaeagnus pungens*

Zones 6 7 8 9 10

Evergreen spreading shrub, growing to 8-15′ in 8-10 years. Easily transplanted.

Flowers in late fall, small fragrant ivory-white blossoms.

Fruit mature by spring, ½″ red berries.

Leaves 2-4″ long with undulating margins, dark green color all year; silvery-tan undersides.

Soil: almost any, wet or dry.

Light: full sun or light shade.

TOLERATES heat and wind.

Prune at any time; shear hedge or clip out individual stems.

Uses: screen, hedge, in shrub borders.

EUONYMUS HEDGE *Euonymus fortunei 'Berryhill'*

Zones 5 6 7 8 9 10

Evergreen shrub, more upright than spreading. Height about 2½′ in 5 years.

Leaves 1½-2″ long, glossy dark green. Holds leaves through all but severest winters.

Soil: almost any, wet or dry.

Light: full sun or light shade.

Prune at any time to shape plant. Responds well to shearing.

Uses: low hedge, plant 12-18″ apart, foundation planting.

CONVEXLEAF HOLLY *Ilex crenata 'Convexa'*

Zones 5 6 7 8 9 10

Low spreading evergreen shrub. Height 3-4′ in about ten years, spreads to 6″ if unpruned. Inconspicuous flowers and black fruit.

Leaves: ½-1″ long, convex, glossy dark green.

Soil: Almost any, prefers slightly acid; tolerates moist position. Mulch rather than cultivate around plants.

Light: Best in full sun.

Pruning: Rarely needed. Trim hedge in summer.

Uses: Low hedges, ideal for edging or below windows.

LITTLE LEAF JAPANESE HOLLY *Ilex crenata 'Microphylla'*

Zones 6 7 8 9 10

Stiff compact evergreen shrub. Height to about 6′ in 10 years if unpruned. Inconspicuous flowers, black fruit.

Leaves: ½-1″ long, finely toothed, dark green.

Soil: almost any, prefers slightly acid conditions. Tolerates moist position. Mulch rather than cultivate around plants.

Light: full sun or light shade.

Pruning rarely needed; shear or trim to size.

Uses: hedge or screen.

AMUR PRIVET *Ligustrum amurense*

Zones 3 4 5 6 7 8 9 (not southern Florida) & 10.

Densely branching deciduous shrub. Upright growth, height to 12-15′ if unpruned.

Flowers in June, small clusters of white blossoms, scented.

Fruit: ¼″ black berries, persist, seldom eaten by birds.

Leaves: 1-3″ tough smooth ovals, dark green, hold to late fall.

Soil: almost any, well drained. Heavy feeder; add peatmoss, compost or leafmold.

TOLERATES urban pollution.

Light: full sun or light shade.

Pruning: clip or trim regularly as hedge. Remove dead and older branches in early spring, cutting right to the base to maintain succession of new strong growth.

Uses: hedge or screen, planted at 12-18″. Each spring cut back new growth to half until desired height is reached.

WAX-LEAF PRIVET *Ligustrum japonicum 'Texanum'*

Zones 7 8 9 10

Fast growing evergreen shrub, also grown as single trunk tree. Height reaches 30-40′ in 15-30 years if unpruned.

Flowers in August, 8-9″ clusters of pungent white blossoms.

Fruit: prolific clusters of small black persistent berries.

Leaves 3-6″ long, pointed, medium green color all year.

Soil: almost any, well drained. Privets are heavy feeders: add peatmoss, compost or leafmold to soil.

Light: full sun or light shade.

TOLERATES urban pollution.

Pruning: clip or trim regularly as hedge. Remove dead and older wood in early spring, cutting right back to base to maintain succession of strong new growth.

Uses: hedge or screen, plant 18-36″ apart; cut back new growth to half each spring until desired height is reached. Shrub borders, foundation plantings, evergreen tree (single trunk).

GOLDEN VICARY PRIVET *Ligustrum ×vicaryi*

Zones 4 5 6 7 8 9
(not southern Florida) & 10

Densely branching deciduous shrub. Unpruned height reaches 10-12′.

Flowers in summer, small clusters of white blossoms, scented.

Fruit: ¼″ black berries, persist, seldom eaten by birds.

Leaves: 1-3″ long, smooth ovals, bright greenish gold throughout summer. Leaves hold to late fall.

Soil: almost any, well drained. Heavy feeder; add peatmoss, compost or leafmold.

Light: full sun or light shade.

TOLERATES urban pollution.

Pruning: clip or trim regularly as hedge. Remove dead or older branches in early spring, cutting right to base to maintain succession of new strong growth.

Uses: hedge or screen, planted at 12-18″. Each spring, cut back new growth to half until desired height is reached.

TATARIAN HONEYSUCKLE *Lonicera tatarica*

Zones 4 5 6 7 8 9
(not southern Florida) & 10.

Upright deciduous shrub growing to 8-10′ tall.

Flowers in late spring, varieties offer color range white, pink or red. Red berries last until fall.

Leaves dark green ovals with bluish undersides.

Soil: almost any, moist and well drained.

Light: best in full sun; tolerates partial shade.

TOLERATES urban pollution, poor soil, insect pests.

Pruning rarely needed except to remove dead wood and older thick stems. Trim as required for shape and size.

Uses: specimen, hedge.

EASTERN WHITE PINE *Pinus strobus*

Zones 4 5 6 7 8
and north in zone 10

Fast growing pyramid when young, loses lower branches with age. Height 18-25′ in 15 years, matures to 75-130′.

Needles in groups of five, soft and blue-green.

Soil: almost any, well drained; tolerates poor, rocky, light and sandy conditions.

Light: full sun though needs light shade in extreme heat.

Uses: specimen or accent, screen or windbreak.

LOMBARDY POPLAR *Populus nigra 'Italica'*

All zones

Very fast growing upright columnar deciduous tree with erect close branches. Grows 3-5′ each year when young; an 8-10′ tree grows to 20′ in 3-5 years. Spreads 5-10′. Short lived, very susceptible to canker. Use for short term shelter belt or windbreak. Plant female trees only.

Leaves large dark green.

FALL COLOR: golden yellow.

Soil: almost any, deep and well drained. Roots tend to plug up drains, heave up sidewalks and pavement if too near.

Light: best in full sun.

TOLERATES drought.

Prune in summer and fall.

Uses: quick growing temporary screen for maximum time of 25-30 years.

TALLHEDGE *Rhamnus frangula 'Columnaris'*

Zones 3 4 5 6 7 8

Large erect deciduous shrub. Height 10-12′; spread 4′ only.

Flowers in early summer, small greenish yellow blossoms followed by ¼″ berries ripening from green, through red to black in late summer. Birds eat berries and scatter seeds, which readily germinate.

Leaves 1½-3″ long, dark green and glossy.

FALL COLOR: yellow, leaves hold late.

Soil: almost any; tolerates dry or wet conditions.

Light: full sun or light shade.

Prune in early spring to contain growth.

Uses: hedge or screen, plant 2-2½′ apart and cut back once to 18-24″ above ground to encourage strong growth.

ALPINE CURRANT *Ribes alpinum*

Zones 3 4 5 6 7 8

Compact erect deciduous shrub, prune for dense spreading form. Height 2-6′.

Flowers in spring, small greenish-yellow blossoms followed by scarlet fruit.

Leaves early, dark green, color holds long into fall.

FALL COLOR: bronze red.

Soil: almost any, well drained. Apply mulch each spring after soil warmed.

Light: sun or shade.

TOLERATES poor soil, shady position.

Prune in late spring. Cut some older shoots to ground level, to encourage new growth from base. Remove ¼ to ⅓ of the shoots each year for succession of vigorous new growth.

Uses: low hedge 1-4′ high, plant 12-15″ apart and tipped back to 6″. Tip again once or twice in first year, then shear once or twice each season.

DWARF ARCTIC or BASKET WILLOW *Salix purpurea 'Nana' (S. p. 'Gracilis')*

All zones

Deciduous twiggy low growing shrub; natural height 24-36″, can be clipped and held as low as 12″. Purplish bark color. Inconspicuous flowers and fruit.

Leaves very narrow, gray or blue-green color.

Soil: almost any, very tolerant of heavy and wet conditions.

Light: full sun or shade.

Prune at any time to retain formal shape.

Uses: low hedge, plant 8″ apart, color contrast in shrub border, for wet places where little else will grow.

CANADIAN HEMLOCK *Tsuga canadensis*

Zones 4 5 6 7 8
and north in zone 10

Broad pyramid with arching, drooping branches; fast growth rate. Height 12-16′ in 10 years, matures to 80-100′ if left unpruned.

Needles soft and dark green with whitish undersides; color holds all year.

Soil: moist, deep, well drained; prefers acid conditions though will tolerate some lime if peatmoss, leafmold or compost is added.

Light: full sun or shade. Shelter from winds at all times.

Pruning: shear for formal effect; very tolerant of clipping.

Uses: good specimen for dark corners, north side of building hedge or screen.

Vines

CLEMATIS HYBRIDS

Zones 4 5 6 7 8 9 10

Fast growing deciduous vines, natural climbers with twining leaf stems. Will grow to height of 6-15′.

Flowers in early summer on current season's growth. Large flat blossoms may be up to 10″ across; varieties in blue, purple, pink, red or white.

Soil: moist, light, and slightly alkaline; prefer some lime in soil. Shallow rooted plants, best left undisturbed by cultivation. Protect roots and lower stems from heat and damage.

Light shade, cool position for roots. Flowers like some sun, though protect from full heat of midday and afternoon sun.

Prune newly planted clematis hybrids to 6-9″ above the ground. Each spring clean out dead wood and reduce shoot length before growth starts. For large blossoms, cut back to the lowest 2 buds on each shoot.

Uses: on wall, trellis or porch, container plant for patio or balcony use, good at back of shrub borders where roots remain shaded and protected.

JACKMAN CLEMATIS *Clematis hybrids Jackmanii Group*

Zones 4 5 6 7 8 9 10

Fast growing deciduous vine, natural climber with twining leaf stems. Will grow to about 12′.

Flowers in early summer on current season's growth. Large flat blossoms may be up to 10″ across; rich purple color.

Soil: moist, light, and slightly alkaline; prefers some lime in soil. Shallow rooted, best left undisturbed by cultivation. Protect roots and lower stems from heat and damage.

Light shade, cool position for roots. Flowers like some sun, though protect from full heat of midday and afternoon sun.

Prune newly planted clematis to 6-9″ above the ground. Each spring clean out dead wood and reduce shoot length before growth starts. For large blossoms, cut back to the lowest 2 buds on each shoot; this vigorous variety can be cut right back to about 36″.

Uses: on wall, trellis or porch; container plant for patio or balcony; good at back of shrub borders where roots remain shaded and protected.

SWEET AUTUMN CLEMATIS *Clematis paniculata (C. dioscoreifolia robusta)* Zones 4 5 6 7 8 9 10

Fast growing deciduous vine, natural climber with twining leaf stems. Rapidly makes dense growth to 30′ high.

Flowers in late summer, profuse and fragrant white blossoms.

Soil: moist, light and slightly alkaline; prefers some lime in soil. Shallow rooted, best left undisturbed by cultivations. Protect roots and lower stems from heat and damage.

Light shade, cool position for roots. Flowers like some sun, though protect from full heat of midday and afternoon sun.

Prune newly planted clematis to 6-9″ above the ground. Each spring clean out dead wood and reduce shoot length as desired, before growth starts.

Uses: informal vine for climbing on wall, trellis, fence or trees; good at back of shrub border where roots remain shaded.

CARLOLINA JESSAMINE *Gelsemium sempervirens*

Zones 7 8 9 10

Semi-evergreen or evergreen vine, will grow to 10′ high with supporting ties. Without support the plant makes a dense mound with gracefully arching branches.

Flowers from early spring to summer, during winter in sheltered places. Blossoms are light yellow, 1½″ across, and fragrant.

Leaves dark green, in groups of 3 leaflets borne on 4-angled stems.

Soil: almost any, moist yet well drained.

Light: full sun to light shade.

Prune to encourage branching; tip long shoots in spring.

Uses: flowering vine for wall, trellis, porch, support needed; plant on top of bank or terrace for trailing effect; will form mound over rock or tree stump.

HALL'S JAPANESE HONEYSUCKLE *Lonicera japonica 'Halliana'*

Zones 4 5 6 7 8 9 (not southern Florida) & 10.

Vigorous twining vine, deciduous or semi-evergreen.

Flowers pure white, turning yellow with age.

Fruit: black.

Leaves: dull green ovals.

Soil: almost any, moist and well drained.

Light: best in full sun.

TOLERATES urban pollution, insect pests, poor soil.

Prune to contain size as desired; remove older branches.

Uses: hedge, plant at 3′ with support, specimen climber, train over fence for screen or barrier, ground cover, erosion control.

VIRGINIA CREEPER *Parthenocissus quinquefolia (Ampelopsis quinquefolia)* Zones 4 5 6 7 8 9 10

Fast growing deciduous creeping and climbing vine; a natural climber with root-like tendrils which penetrate the support if possible. Side branches fill out growth to give it depth and density. Climbs to about 50′.

Fruit: blue-black berries ripen in fall; eaten by birds.

Leaves compound with 5 fingerlike leaflets, dark green.

FALL COLOR: brilliant scarlet.

Soil: fertile and moist.

Light: full sun or light shade.

Uses: leafy vine for wall or trellis, no supports required once plant established; background for flowering specimens; ground cover for banks or terraces, will climb any fence or tree encountered.

CHINA FLEECE VINE, SILVER LACE VINE *Polygonum aubertii*

Zones 5 6 7 8 9 10

Fast growing deciduous vine with twining stems. Dense growth; may grow as much as 20-30′ in a single year. If top is killed by severe frosts, will grow back from roots and flower in one season.

Flowers throughout summer; dense clusters of fluffy greenish-white blossoms.

Soil: almost any, well drained.

Light: best in full sun.

TOLERANT of drought, poor soil conditions, and resists attack by pests and diseases.

Pruning rarely needed.

Uses: informal vine makes thick growth along fence or on strong trellis.

WISTERIA SINENSIS

Zones 5 6 7 8 9 10 (cultivars suited to different areas)

Deciduous woody vines, natural climbers with twining stems, will climb as high as 40′ if support available. May also be grown as self-supporting specimen trees. Light brown twisted wood gives characteristic aged look in winter.

Flowers in pendulous clusters in spring, often before leaves unfold. Clusters are up to 36″ long; fragrant; varieties have white as well as many hues of violet-blue and purplish flowers.

Leaves compound, up to 19 or more light green leaflets.

Soil: fertile, light, moist yet well drained. Good near water where roots can obtain unlimited water supply. Overfeeding results in too much vegetative growth and no flowers.

Light: best in full sun.

Prune in spring to control height, and train as desired.

Uses: flowering climber on wall (strong supports needed), trellis or arbor. Will grow up tree, though supporting tree may be strangled by wisteria. Self-supporting specimen: stake main stem and prune back top growth severely for several years to encourage main stem to thicken into supportive trunk.

merchants publishing company
20 mills street • kalamazoo, michigan 49001
telephone 616-345-1175

Appendix

TO MAKE THIS BOOK MORE USEFUL, THE FOLLOWING LISTS ARE PROVIDED.

Size Breakdown

BROADLEAF EVERGREEN SHRUBS

Low Growing
(under 3′)

Buxus microphylla var. koreana 'Wintergreen'
Calluna vulgaris
Erica carnea
Ilex cornuta 'Burfordii Nana'
I. c. 'Rotunda'
I. crenata 'Compacta'
I. c. 'Helleri'
I. vomitoria 'Nana'
Mahonia aquifolium 'Compacta'
Santolina chamaecyparissus
S. virens

Medium
(3′-6′)

Abelia ×grandiflora
Aucuba japonica
A. j. 'Variegata'
Azalea—see Rhododendron
Berberis darwinii
Buxus sempervirens 'Wellerii'
Euonymus fortunei
E. f. 'Sarcoxie'
E. f. var. vegeta
E. f. 'Emerald 'n Gold'
E. kiautschovica
Gardenia jasminoides
Ilex crenata 'Convexa Northern Beauty'
Mahonia aquifolium
Nandina domestica
Pieris japonica
Rhododendron indicum—Southern Indica Group
R. obtusum var. kaempferi
R. o. Kurume Group
R. ×kosteranum

Tall
(over 6′)

Elaeagnus pungens
E. p. 'Maculatus'
Euonymus japonica 'Albomarginata'
E. j. 'Aureo-marginata'
Ilex cornuta
I. c. 'Burfordii'
I. c. 'Rotunda'
I. crenata
I. opaca
I. vomitoria
Nerium oleander
Photinia glabra
Pittosporum tobira
P. t. 'Variegata'
Pyracantha coccinea 'Lalandei'
P. fortuneana 'Graberi'
Rhododendron species (includes Azalea)
Viburnum rhytidophyllum

CONIFEROUS EVERGREEN SHRUBS

Low
(under 3′)

Chamaecyparis obtusa 'Nana Gracilis'
Juniperus chinensis 'Pfitzeriana'—Nick's Compact
J. c. 'Procumbens Nana'
J. c. 'P. Variegata'
J. c. 'Sargentii'
J. conferta
J. horizontalis 'Douglasii'
J. h. 'Plumosa'
J. h. 'P. Compacta Youngstown'
J. h. 'Wiltonii'
J. sabina 'Scandia'
J. s. 'Tamariscifolia'

Picea abies 'Nidiformis'
Taxus cuspidata 'Nana'

Medium
(3′-6′)

Chamaecyparis pisifera 'Filifera Aurea'
Juniperus chinensis 'Armstrongii'
J. c. 'Blaauwii'
J. c. 'Hetzii'
J. c. 'Maneyi'
J. c. 'Pfitzeriana'
J. c. 'P. Aurea'
J. c. 'P. Glauca'
J. excelsa 'Stricta'
J. squamata 'Meyeri'
Picea glauca 'Conica'
Pinus mugo
Platycladus orientalis 'Berckmann's' (Thuja orientalis)
Taxus cuspidata 'Densa'
T. ×media 'Brownii'
T. ×m. 'Sebian'
T. ×m. 'Tauntonii'
T. ×m. 'Wardii'
Thuja occidentalis 'Globosa'

Tall
(over 6′)

Juniperus chinensis 'Ames'
J. c. 'Keteleeri'
J. c. 'Mas'
J. c. 'Torulosa'
J. communis 'Stricta'
J. scopulorum 'Blue Haven' or 'Blue Heaven'
J. s. 'Moffetti'
J. s. 'Moonglo'
J. virginiana 'Burkii'
J. v. 'Canaerti'
J. v. 'Hillii'
J. v. 'Skyrocket'
Taxus cuspidata
T. c. 'Capitata'
T. ×media 'Hicksii'
Thuja occidentalis 'Nigra'
T. o. 'Pyramidalis'
T. o. 'Rheingold'
T. o. 'Techny'
T. o. 'Wareana'

CONIFEROUS EVERGREEN TREES

Medium
(20′-35′)

Picea pungens 'Moerheimii'
Pinus cembra
Pinus pinea
Podocarpus macrophyllus var. maki

Tall
(over 35′)

Abies concolor
Cedrus atlantica
C. deodara
Cupressus sempervirens 'Glauca'
Picea abies
P. engelmannii 'Glauca'
P. glauca
P. g. 'Densata'
P. mariana
P. pungens
P. p. 'Glauca'
Pinus caribaea
P. halepensis
P. nigra
P. ponderosa
P. radiata
P. resinosa
P. strobus
P. sylvestris
Podocarpus macrophyllus
Pseudotsuga menziesii
Tsuga canadensis

DECIDUOUS SHRUBS

Low
(under 3′)

Berberis thunbergii 'Atropurpurea Nana'
Hypericum frondosum
Potentilla fruticosa
Rosa hybrids
Spiraea ×bumalda 'Anthony Waterer'
Viburnum opulus 'Nanum'

Medium
(3′-6′)

Berberis thunbergii 'Atropurpurea'
Cotoneaster apiculatus
Cytisus ×praecox
Deutzia gracilis
D. ×magnifica
Euonymus alata 'Compacta'
Hydrangea arborescens 'Grandiflora'
H. a. 'G. Annabelle'
H. macrophylla
Kerria japonica 'Pleniflora'
Ligustrum ×vicaryi
Lonicera xylosteum 'Claveyi'
Prunus glandulosa
Rosa hybrids
Spiraea prunifolia
S. ×vanhouttei
Syringa ×persica

Tall
(over 6′)

Acer ginnala
A. palmatum 'Atropurpureum'
Berberis thunbergii
B. t. 'Aurea'
Carangana arborescens
Chaenomeles speciosa
Cornus alba 'Argenteo-marginata'
Cotinus coggygria
C. c. 'Royal Purple'
Cotoneaster acutifolia
C. multiflorus
Euonymus alata 'Monstrosa'
E. europaea
Forsythia ×intermedia
Hibiscus syriacus (Althea syriacus)
Hydrangea paniculata 'Grandiflora'
Ilex verticillata
Kolkwitzia amabilis
Lagerstroemia indica
Lonicera fragrantissima
L. tatarica
Magnolia stellata
Philadelphus coronarius 'Aureus'
P. ×virginalis
Physocarpus opulifolius 'Luteus'
Prunus ×cistena
Rhododendron spp. (includes Azalea)
Sambucus canadensis
Sorbus ×tianshanica
Syringa vulgaris
Tamarix ramosissima (T. pentandra)
Viburnum carlcephalum
V. dentatum
V. lantana
V. opulus 'Roseum'
V. plicatum var. tomentosum
V. sieboldii
Vitex agnus-castus
Weigela florida and hybrids

FLOWERING TREES

Small
(under 20′)

Franklinia alatamaha
Magnolia ×soulangiana
Malus hybrid 'Almey'
M. hybrid 'Royalty'
M. hybrid 'Vanguard'
M. sargentii
M. s. 'Rosea'
M. scheideckeri

Medium
(20′-35′)

Aesculus ×carnea 'Briottii'
Albizia julibrissin
Amelanchier canadensis
Cercis canadensis
Chionanthus virginicus
Cornus florida
C. f. 'Rubra'
C. kousa
Crataegus laevigata (C. oxycantha)
C. crus-galli
C. phaenopyrum (c. cordata)
Koelreuteria paniculata
Malus hybrid 'Dolgo'
M. hybrid 'Dorothea'
M. hybrid 'Hopa'
M. hybrid 'Radiant'
M. hybrid 'Red Silver'
M. purpurea 'Eleyi'
Prunus cerasifera 'Pissardi'
P. persica

P. serrulata 'Kwanzan'
P. subhirtella 'Pendula'
Sorbus aucuparia

Tall
(over 35′)

Liriodendron tulipifera
Magnolia grandiflora

SHADE TREES

Medium
(20′-35′)

Celtis occidentalis
Elaeagnus angustifolia

Tall
(over 35′)

Acer platanoides
A. p. 'Columnare'
A. p. 'Crimson King'
A. p. 'Schwedleri'
A. rubrum
A. saccharinum
A. saccharum
Betula nigra
B. papyrifera
B. pendula
B. p. 'Dalecarlica'
Cercidiphyllum japonicum
Fraxinus pennsylvanica
Ginkgo biloba
Gleditsia triacanthos var. inermis
G. t. var. i. 'Sunburst'
Juglans nigra
Liquidambar styraciflua
Morus alba—Fruitless
Platanus occidentalis
Populus nigra 'Italica'
Quercus palustris
Salix babylonica
S. matsudana 'Tortuosa'
Tilia cordata
T. ×euchlora
Ulmus parvifolia

HEDGES & SCREENS

Low
(under 3′)

Berberis thunbergii 'Atropurpurea Nana'
Euonymus fortunei 'Berryhill'
Ribes alpinum
Salix purpurea 'Nana'

Medium
(3′-6′)

Buxus microphylla var. japonica
Ilex crenata 'Convexa'
I. c. 'Microphylla'
Ligustrum ×vicaryi

Tall
(over 6′)

Elaeagnus pungens
Ligustrum amurense
L. japonicum 'Texanum'
Lonicera tatarica
Rhamnus frangula 'Columnaris'
Tsuga canadensis

WINDBREAK TREES

Tall
(over 35′)

Populus nigra 'Italica'
Pinus strobus
Tsuga canadensis

Index

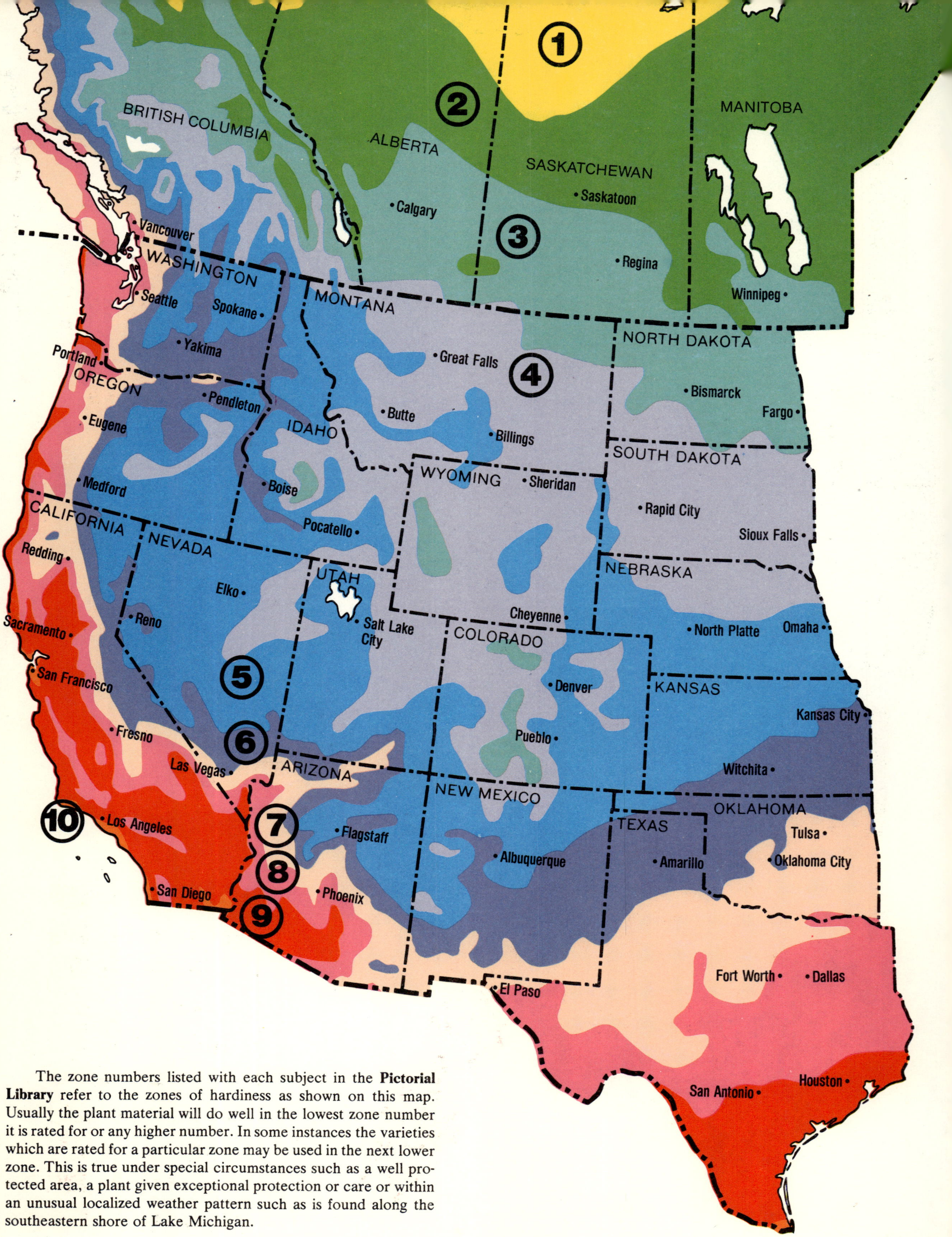

The zone numbers listed with each subject in the **Pictorial Library** refer to the zones of hardiness as shown on this map. Usually the plant material will do well in the lowest zone number it is rated for or any higher number. In some instances the varieties which are rated for a particular zone may be used in the next lower zone. This is true under special circumstances such as a well protected area, a plant given exceptional protection or care or within an unusual localized weather pattern such as is found along the southeastern shore of Lake Michigan.